A COURSE I

A Course in
Basic Scientific English

J. R. Ewer Department of English, University of Chile
and
G. Latorre School of Engineering, University of Chile

Longman

LONGMAN GROUP LIMITED
London

Associated companies, branches and representatives throughout the world

First published 1969
*New impressions *1970; *1971; *1974; *1975*

ISBN 0 582 52009 6

Printed in Hong Kong by
Dai Nippon Printing Co. (Hong Kong) Ltd.

Acknowledgements

This approach to the problem of preparing a special English course for students of science owes much to the writings of the pioneers in this still relatively undeveloped field, notably Dr D. J. Bruton, Dr W. E. Flood, Dr C. L. Barber, Mr D. Y. Morgan and, in particular, Dr Michael West, whose work is always a source of inspiration for the practical teacher.

We are also most grateful to those of our colleagues, both inside and outside the University of Chile, who have been using the materials in the classroom and whose comments have made this book better than it would otherwise have been.

We are grateful to the following for permission to reproduce copyright material:

W. H. Freeman & Co Inc for an extract from 'Food' by Nevin S. Scrimshaw from *Scientific American* issue dated September 1963, and an extract from 'Chelation in Medicine' by Jack Schubert from *Scientific American* issue dated May 1966. George G. Harrap & Co Ltd and Doubleday & Co Inc for an extract from *Research Problems in Biology*, Series 4, Copyright © 1965 by Regents of the University of Colorado; authors, and the Proprietors of *The Listener* for an extract from an article by H. D. Dunn from issue dated 3rd February 1966, and an extract from an article by M. Hammerton from issue dated 18th October 1962; Longmans, Green & Co for an extract from *The Tools of Social Science* by J. Madge; McGraw-Hill Co Ltd for an extract from *Engineering as a Career* by R. J. Smith, and an extract from *Experimental Chemistry* by Sienko and Plane; Thomas Nelson & Sons Ltd for an extract from *Principles of Physical Geology* by Arthur Holmes; the Proprietors of *The Observer* for an extract from an article by John Davy from issue dated 19th December 1965; Penguin Books Ltd for an extract from Sir Willis Jackson's article in *Penguin Technology Survey 1966*, an extract from J. R. Busvine's article in *Penguin Science Survey 1966*, and an extract from *Fundamentals of Psychology* by C. J. Adcock; Routledge & Kegan Paul Ltd for an extract 'Sociology' by M. D. King from *The Social Sciences* edited by D. C. Marsh; the Proprietors of *Think* for an extract from an article by W. Weaver from issue dated April 1961, Copyright © 1961 by Warren Weaver, published by I.B.M.; UNESCO Courier for an extract from 'A World-wide programme of Scientific Research' by Michel Batisse, from issue dated July–August 1964; United Nations Organization for an adaptation from the *Interim Report on the United Nations Development Decade* (document E/4196 of 5th May 1966) prepared by the Secretary General on behalf of the Administrative Committee on Co-ordination, and an extract from a *Report on the United Nations Conference on the Application of Science and Technology for the benefit of the less developed areas*, Copyright © 1963 the United Nations; and the United States Atomic Energy Commission for an extract from *Offsite Ecological Research of the Division of Biology and Medicine*.

Contents

Introduction

PURPOSE AND SCOPE OF THE COURSE

Purpose

The purpose of this course, as its title indicates, is to teach students of scientific subjects (including medicine, engineering and agriculture) the basic language of scientific English. This basic language is made up of sentence patterns, structural (functional) words and non-structural vocabulary which are common to all scientific disciplines and form the essential framework upon which the special vocabulary of each discipline is superimposed. Once this basic language has been mastered—together with the principal word-building devices (prefixes and suffixes) also presented in this book—the acquisition of these special vocabularies presents very little difficulty, since they are mainly *international* words and therefore very similar to those already used in the student's own language.

Nature of the linguistic material presented

The material incorporated in the course has been selected, for the most part on a frequency basis, from the scrutiny of more than three million words of modern scientific English of both American and British origin.[1] This sample covered ten broad areas of science and technology (physics, chemistry, biology, geology and geomorphology, medicine, engineering, sociology, economics, psychology and agriculture) and represented the types of literature likely to be consulted by students or graduates of science—university textbooks, professional papers and articles, scientific dictionaries and semi-popularizations. Whilst the principal criteria for the inclusion of items were frequency and range, a certain amount of material was selected for other reasons, e.g. because of their usefulness as describers or definers, because they were members of a group or set, or because, though not unduly frequent, they were essential or non-substitutable (as is the case with the Present Continuous tense, for example).[2]

Grading and flexibility

Although it is assumed that students using this course have already received a certain amount of training in English at school or in a language institute, the material included has, in its presentation, been graded in length and complexity. Hence the most frequently used and simple structures have been introduced first, the whole of the corresponding Unit being written as far as possible exclusively in terms of the structure being presented (thus, for example, all the verbs appearing in Unit I are in the Present Simple Tense, which is the main structure in this unit). The length of the reading passages—and therefore of the amount of material they contain—increases progressively, from about 450 words in the early units to nearly three times

[1] It should be stated that, as far as scientific English is concerned, no significant difference was found between these two varieties of English. The few minor points of variance occurring in the course itself are explained.

[2] Further details are given in *English Language Teaching*, Vol. XXI, No. 3.

this length in later ones. There is a fair amount of repetition of phrasing within each individual unit and this repetition is carried over to the exercise sections of the following unit or two, so that revision of the immediately preceding work goes on all the time. The special Revision Units (Nos. 5 and 10, revising the material presented in Units 1–4 and 6–9 respectively, and the General Revision Units Nos. 11 and 12, reviewing the principal elements of the course as a whole) will, it is hoped, constitute a valuable aid in this very necessary task of consolidation.

Since individual students and even whole classes of students may show a good deal of variation with regard to the knowledge of English they bring to the course or the speed with which they work, an effort has been made to make the course flexible enough to cover these contingencies. This has been done by giving a wide choice of exercises in the Word Study and Structure Study sections, by incorporating additional suggestions for exercises in the Teacher's Notes, by including a Discussion and Criticism section designed to give students the opportunity to produce continuous English based on the contents of the reading passages, and by adding a Supplement of extracts from the original literature of modern science.

The oral approach

The approach used throughout the book is essentially an oral one, in view of the fact that: (a) oral repetition (in context) is the most effective way of fixing material, even for purely recognition purposes; (b) much more work can be accomplished orally than in written form, over a given period; (c) oral work adds variety and interest to the lessons. As many teachers will be aware, there are also two additional factors operating in various parts of the world which lend urgency to this emphasis upon oral work —firstly, the number of English-speaking specialists visiting non-English-speaking countries to give lecture-cycles or direct seminars is increasing rapidly, and the widespread failure of students or local specialists to understand oral scientific English, and be able to communicate themselves, is robbing these countries of much of the value which would otherwise be gained from these visits; secondly, in spite of the growing number of scholarships and fellowships to English-speaking countries, many first-rate science students are failing to obtain them because of their lack of knowledge of the language. This again is a serious loss of human resources.

Further aims of the course

In addition to the purely language-teaching aims outlined above, the course is designed to stimulate critical thought and foster the habits of clear exposition and the impartial examination of evidence; at the same time an attempt has been made to encourage students to take an active interest in their own discipline and its relationships with other sciences and with society as a whole. Thus it is hoped that it will serve a broadly educational purpose as well as its specific linguistic one.

LIBRARY SUGGESTIONS

The effectiveness of the course will be much increased if an adequate reference library can be provided. Some suggestions—mainly referring to inexpensive paperbacked editions—are as follows:

Language

1 Dictionaries

FLOOD & WEST *Elementary Scientific and Technical Dictionary*, Longmans
(explains over 10,000 scientific terms within a vocabulary of less than 2000 words; illustrated)
The English Duden, Bibliographisches Institut, Mannheim, and Harrap
(very complete picture-dictionary which includes the main sciences and many branches of engineering)
Chambers Technical Dictionary, Chambers
(available in English-English or multilingual form)

2 Pronunciation Drilling

CLAREY & DIXSON *Pronunciation Exercises in English*, Regents Publishing Co.

Science

1 General Science and the Scientific Attitude

PYKE *The Boundaries of Science*, Penguin
BECK *The Simplicity of Science*, Penguin
CONANT *Science and Common Sense*, Yale Paperbacks
CANNON *The Way of an Investigator*, Norton
HILDEBRAND *Science in the Making*, Columbia Paperbacks
EDGE (ed.) Experiment: a series of scientific case-histories, B.B.C.
NEWMAN (ed.) *What is Science?* Washington Square

2 Mathematics

PEDOE *The Gentle Art of Mathematics*, Penguin
SAWYER *Mathematician's Delight*, Penguin
ADLER *The New Mathematics*, Signet
SUTTON *Mathematics in Action*, Harper Torchbooks

3 Statistics

WALLIS & ROBERTS *The Nature of Statistics*, Collier
HUFF *How to Lie with Statistics*, Gollancz
REICHMANN *The Use and Abuse of Statistics*, Penguin

4 Physics

ANDRADE *Physics for the Modern World*, Barnes & Noble
ISAACS *Introducing Science*, Penguin
BITTER *The Education of a Physicist*, Heinemann
BEISER *Physics for Everybody*, Everyman

5 Chemistry

PORTER *Chemistry for the Modern World*, Barnes & Noble
LESSING *Understanding Chemistry*, Signet
HUTTON *Chemistry*, Penguin
JAFFE *Chemistry Creates a New World*, Pyramid

6 Earth Sciences

DURY *The Face of the Earth*, Penguin
CRONEIS & KRUMBEIN *Down to Earth*, Phoenix
SWINNERTON *The Earth Beneath Us*, Penguin
RAPPORT *The Crust of the Earth*, Signet

7 Social Sciences

SIMPSON *Man in Society*, Random House
KLUCKHOHN *Mirror for Man*, Premier
WRIGHT MILLS *The Sociological Imagination*, Evergreen Books
KARDINER *They Studied Man*, Mentor
LIPSET *Political Man*, Anchor
MEAD (ed.) *Cultural Patterns and Technical Change*, Mentor
BERELSON (ed.) *The Behavioral Sciences Today*, Harper Torchbooks

8 Economics

SOULE *Ideas of the Great Economists*, Mentor
ROSTOW *The Stages of Economic Growth*, Cambridge Paperbacks
ROBINSON *Economic Philosophy*, Penguin
THEOBALD *The Rich and the Poor*, Mentor

9 Biology and Biochemistry

WADDINGTON *Biology for the Modern World*, Barnes & Noble
WORTH & ENDERS *The Nature of Living Things*, Signet
WINOKER *General Biology*, Littlefield
ASIMOV *The Chemicals of Life*, Signet

10 Medicine

MARGERSON *Medicine Today*, Penguin
CALDER *Medicine and Man*, Signet
ATKINSON *Magic, Myth and Medicine*, Premier
BROCKINGTON *World Health*, Penguin

11 Psychology

ADCOCK *Fundamentals of Psychology*, Penguin
EYSENCK *Uses and Abuses of Psychology*, Penguin
CROW *Readings in General Psychology*, College Outlines
CROW *Outline of General Psychology*, Littlefield

12 Engineering

(Note: Paperbacks on Engineering are very scarce: the few listed below are, however, very good)

WILLIAMS & FORBES *Building an Engineering Career*, McGraw-Hill

FINCH *The Story of Engineering*, Anchor

CROSS *Engineers and Ivory Towers*, McGraw-Hill

13 Agriculture

(As for Engineering, paperbacks on Agriculture are practically non-existent. The following may provide useful related reading)

FISHWICK *Teach Yourself Farming*, English Universities Press

STORER *The Web of Life*, Signet

It is further suggested that lists of publications on agriculture should be obtained from:

(a) U.S. Department of Agriculture, Washington, U.S.A.

(b) Her Majesty's Stationery Office, London, England, as the bulletins, etc. issued are varied and cheap.

14 Journalism

(Students of journalism should, of course, be encouraged to read in all branches of science; the books listed under 'General Science' above are particularly suitable. Some more specialized books are listed below)

UNESCO *The Training of Journalists*, UNESCO

THOULESS *Straight and Crooked Thinking*, Pan Books

LEDERER *A Nation of Sheep*, Crest Books

GOLDWIN *Towards the Liberally Educated Executive*, Mentor

WILLIAMS *Communications*, Penguin

HOGGART *The Uses of Literacy*, Penguin

THOMPSON *Discrimination and Popular Culture*, Penguin

15 History of Science

FORBES *A History of Science & Technology*, Vol. 2, Penguin

HALL *A Brief History of Science*, Signet

CRANE *Giants of Science*, Pyramid

PLEDGE *Science since 1500*, Harper

MASON *A History of the Sciences*, Collier

16 Experimental Methods

BEVERIDGE *The Art of Scientific Investigation*, Vintage Paperbacks

WILSON *Introduction to Scientific Research*, McGraw-Hill

17 Instruments

ADLER *The Tools of Science*, Day

18 Science and Society

RUSSELL *The Impact of Science on Society*, Simon & Schuster
SNOW *The Two Cultures*, Mentor
BERKNER *The Scientific Age*, Yale

19 Report-writing

COOPER *Writing Technical Reports*, Penguin

20 Careers in Science & Technology

GOLDSMITH *Careers in Technology*, Penguin

21 Bibliography of Scientific Paperbacks

Teachers who wish to extend their reference library or compile more specialized reading-lists of low-priced books should consult:

DEASON *A Guide to Science Reading*, Signet
Paperbacks in Print, Whitaker

Unit 1

THE SCIENTIFIC ATTITUDE

What is the nature of the scientific attitude, the attitude of the man or woman who studies and applies physics, biology, chemistry, geology, engineering, medicine or any other science?

We all know that science plays an important role in the societies in which we live. Many people believe, however, that our progress depends on two different aspects of science. The first of these is the application of the machines, products and systems of applied knowledge that scientists and technologists develop. Through technology, science improves the structure of society and helps man to gain increasing control over his environment. New fibres and drugs, faster and safer means of transport,[1] new systems of applied knowledge (psychiatry, operational research, etc.) are some examples of this aspect of science.

The second aspect is the application by all members of society, from the government official to the ordinary citizen, of the special methods of thought and action that scientists use in their work.

What are these special methods of thinking and acting? First of all, it seems that a successful scientist is full of curiosity—he wants to find out how and why the universe works. He usually directs his attention towards problems which he notices have no satisfactory explanation, and his curiosity makes him look for underlying relationships even if the data available seem to be unconnected. Moreover, he thinks he can improve the existing conditions, whether of pure or applied knowledge, and enjoys trying to solve the problems which this involves.

He is a good observer, accurate, patient and objective and applies persistent and logical thought to the observations he makes. He utilizes the facts he observes to the fullest extent. For example, trained observers obtain a very large amount of information about a star (e.g. distance, mass, velocity, size, etc.) mainly from the accurate analysis of the simple lines that appear in a spectrum.

He is sceptical—he does not accept statements which are not based on the most complete evidence available—and therefore rejects authority as the sole basis for truth. Scientists always check statements and make experiments carefully and objectively to verify them.

Furthermore, he is not only critical of the work of others, but also of his own, since he knows that man is the least reliable of scientific instruments and that a number of factors tend to disturb impartial and objective investigation (see Unit 8).

Lastly, he is highly imaginative since he often has to look for relationships in data which are not only complex but also frequently incomplete. Furthermore, he needs imagination if he wants to make hypotheses of how processes work and how events take place.

These seem to be some of the ways in which a successful scientist or technologist thinks and acts.

[1] *transportation* in U.S.A.

Comprehension

1 Name some sciences.
2 Name two ways in which science can help society to develop.
3 Give some examples of the ways in which science influences everyday life.
4 What elements of science can the ordinary citizen use in order to help his society to develop?
5 How can you describe a person who wants to find out how and why the universe works?
6 What is the role of curiosity in the work of a scientist?
7 Name some of the qualities of a good observer.
8 Give an example of how observed facts are utilized to the fullest.
9 How does a sceptical person act?
10 How does the scientist act towards (a) evidence presented by other people, (b) evidence which he presents in his own work?
11 What do you know about the data which the scientist often has to use? How does this affect his way of thinking?
12 For what other purposes does a scientist need imagination?

Word Study
WORD-BUILDING

A common way of making new words in English is by adding standard combinations of letters to existing words, either at the beginning (prefixes) or at the end (suffixes). By noting these carefully, you will find it is easy to make large increases in your recognition vocabulary.

1 The suffix *-ist*

A person who studies and applies
- geology is a geolog*ist*
- biology is a biolog*ist*
- sociology is a ......
- ...... is a chem*ist*
- anthropology is a ......
- ...... is a psycholog*ist*
- archaeology is a ......
- ...... is a ecolog*ist*
- agronomy is a ......

2 The suffix *-(i)an*

A person who studies and applies
- mathematics is a mathematic*ian*
- statistics is a ......
- ...... is an obstetric*ian*

But

A person who applies the study of
- economics is an economist
- engineering is an engineer
- architecture is an architect
- medicine is a doctor[1]

[1] Usually *physician* in U.S.A.

3 The suffix *-ion*

This suffix converts a verb into the corresponding noun. The following are some examples which occur in our first passage:

VERB	NOUN
to act	act*ion*
to apply	applicat*ion*
to observe	observat*ion*

More examples of this suffix are given in the Word Study section of Unit 2.

EXERCISE (a) Form nouns from the following verbs:

to imagine; to attract; to direct; to construct; to connect; to relate; to fluctuate.

(b) Form verbs from the following nouns:

conversion; suggestion; production; definition; operation; reduction; population.

NOTE: to join—junction; to destroy—destruction; to query—question; to transmit—transmission.

4 The prefixes *in-* and *un-*

These prefixes are used to make an adjective negative, e.g. '*in*complete' (l. 45) means 'not complete'; '*un*connected' (l. 24) means 'not connected'.

EXERCISE (a) Using *in-*, make the following negative:

accurate; capable; direct; essential; frequent.

(b) Using *un-*, make the following negative:

able; stable; usual; critical; reliable; successful; imaginative; true.

Structure Study

SIMPLE PRESENT TENSE

The main structure in the passage is the Simple Present Tense. Remember that this tense is used:

(i) for actions in the present which happen usually, habitually or generally, e.g. 'He usually *directs* his attention towards problems which he notices have no satisfactory explanation' (ll. 20–21);

(ii) for stating general truths, e.g. 'science *plays* an important role in the societies in which we live' (ll. 4–5); or for stating scientific laws, e.g. Water *freezes* at 0°C.;

(iii) for describing processes in a general way, e.g. A scientist *observes* carefully, *applies* logical thought to his observations, *tries* to find relationships in data, etc.

EXERCISE (a) Fill in the blanks in the following and repeat aloud several times:

I make		... check	
They ...			
She ...	accurate	... check	the validity of
The scientist ...	experiments	... checks	statements
Scientists ...		... check	
We ...		... check	
You ...			

I think		... observes	
He ...		... observe	
They ...	logically	... observes	accurately
We ...		... observe	
She ...		... observes	
You ...			

(b) Add as many verbs and appropriate complements as possible, chosen from the passage and the Word Study section, to the following subjects: the scientist, scientists, we,

e.g. The scientist	USES	
Scientists	USE	reliable instruments
We	USE	

(c) Repeat Exercise (b) above using the same set of verbs and complements, but using new subjects chosen from the passage or the Word Study section, e.g. *Physicists* use reliable instruments.

The Negative

The Simple Present Tense forms the negative by the use of *do not* or *does not* before the main verb, e.g.

I, you *do not*	
He, she *does not*	KNOW the importance of science.
We, they *do not*	

EXERCISE (d) Fill in the blanks in the following and repeat aloud:

I do not accept	
You ... not accept	incomplete evidence
We ... not accept	unreliable information
A scientist ... not accept	inaccurate statements
They ... not accept	authority in science

(e) Repeat Exercise (a) above, using the negative.

The Interrogative

The Simple Present Tense forms questions by the use of *do* or *does* before the subject of the main verb, e.g.

Do	I you	
Does	he she	KNOW the importance of science?
Do	we they	

EXERCISE (f) Repeat Exercise (d) above, using the question form.

(g) Put the verbs in brackets into their correct forms:

1 A statistician (apply) mathematics in his work.
2 You (accept) incomplete evidence?
3 The evidence (seem) incomplete.
4 The government official (use) objective methods?
5 Trained observers usually (utilize) data to the fullest.
6 He always (try) to look for underlying relationships in collections of data.
7 A scientist always (think) logically?

SUBSTITUTION TABLES
Simple Present Active

A Affirmatives

1	2	3	4	5	6	7
A scientist A technologist A researcher An investigator	often	uses employs needs	mathematics complex instruments imagination statistical methods new apparatus	in	his	work
They Scientists You Researchers		use employ need			their	

B Negatives

1	2	3	4
A physicist A biologist He An engineer	does not	use employ apply	unreliable instruments inaccurate observation unsuccessful techniques
Scientific workers I We Biochemists	do not		

C Questions

1	2	3	4	5	6
Does	a specialist an agronomist he a medical worker	some- times	develop require	new	instruments? techniques?
Do	mathematicians geologists they psychologists		need use		methods? ideas?

Discussion and Criticism

1 Do you think there are other special ways of thinking and acting, used by scientists? If so, comment and explain.

2 Do you think some of these ways are more important than others? If so, give reasons.

3 Do you know of any famous scientist whose work demonstrates some or all the qualities mentioned in the passage? Give details.

4 Try to say something about the work of some of the scientists mentioned in the Word Study section.

5 In what ways do other sciences affect the particular science you study yourself? Give examples.

6 Do you agree that it is important to train the non-scientist to think in a scientific way (ll. 14–17). Give good evidence for your point of view.

7 Do you agree that 'man is the least reliable of scientific instruments' (ll. 40–41)? Give examples.

8 Give a clear explanation of what you think the word 'authority' (l. 36) means.

BIBLIOGRAPHY

CANNON *The Way of an Investigator*, Norton.
WILLIAMS-ELLIS *Modern Scientists at Work*, Harrap.
BURLINGAME *Scientists behind the Inventors*, Avon Books.
CROSS *Engineers and Ivory Towers*, McGraw-Hill Paperbacks.

Unit 2

NUMBERS AND MATHEMATICS

It is said that mathematics is the base of all other sciences, and that arithmetic, the science of numbers, is the base of mathematics. Numbers consist of whole numbers (integers) which are formed by the digits 0, 1, 2, 3, 4, 5, 6, 7, 8 and 9 and by combinations of them. For example, 247—two hundred and forty seven[1]—is a number formed by three digits. Parts of numbers smaller than 1 are sometimes expressed in terms of fractions, but in scientific usage they are given as decimals. This is because it is easier to perform the various mathematical operations if decimals are used instead of fractions. The main operations are: to add, subtract, multiply and divide; to square, cube or raise to any other power; to take a square, cube or any other root and to find a ratio or proportion between pairs of numbers or a series of numbers. Thus, the decimal, or ten-scale, system is used for scientific purposes throughout the world, even in countries whose national systems of weights and measurements are based upon other scales. The other scale in general use nowadays is the binary, or two-scale, in which numbers are expressed by combinations of only two digits, 0 and 1. Thus, in the binary scale, 2 is expressed as 010, 3 is given as 011, 4 is represented as 100, etc. This scale is perfectly adapted to the 'off-on' pulses of electricity, so it is widely used in electronic computers: because of its simplicity it is often called 'the lazy schoolboy's dream'!

Other branches of mathematics such as algebra and geometry are also extensively used in many sciences and even in some areas of philosophy. More specialized extensions, such as probability theory and group theory, are now applied to an increasing range of activities, from economics and the design of experiments to war and politics. Finally, a knowledge of statistics is required by every type of scientist for the analysis of data. Moreover, even an elementary knowledge of this branch of mathematics is sufficient to enable the journalist to avoid misleading his readers, or the ordinary citizen to detect the attempts which are constantly made to deceive him.

Comprehension

1 What is the relationship of mathematics to the other sciences?
2 What is the science of numbers called?
3 Name a two-digit integer.
4 Name two ways of expressing parts of the number *one* (unity).
5 Name the common arithmetical operations. Using actual numbers, give examples of each.
6 What are the two number-systems commonly used throughout the world?
7 Give examples of numbers in the binary system.
8 What are the advantages of each system?
9 Name some other branches of mathematics.

[1] in American usage the *and* is omitted.

10 What branch of mathematics is very useful to the ordinary citizen? Why?

Word Study

SYNONYMS EXERCISE

Find words in the passage which mean approximately the same as:
entire (w...e); usually (f...y); in the place of (i...d of); system of measurement (s...e); widely (ex...y); be put to use in (be a...d to); lastly (f...y); kind, sort (t...e); simpler (e...r); cause someone to make a mistake by giving wrong or incomplete information (m...d); continually (c...y); discover, find out (d...t); action of trying to do something (a...t); a group of measurements, etc. arranged in an orderly way to form a whole (s...m).

WORD-BUILDING

1 The suffix *-ion* (*-ation*, *-ition*)

This suffix forms nouns from verbs with the meaning of: process or result of doing something. Thus *operation* (l. 10) means: process or result of operating. Other nouns formed in this way are: 'add*ition*' (process or result of adding) from (*to*) *add*, 'subtr*action*' from *subtract*, 'di*vision*' from *divide*, 'multipl*ication*' from *multiply*.

EXERCISE

Using *-ation*, make nouns from the following verbs: apply; adapt; specialize; compute; calculate; isolate; combine; explain; investigate.

2 The suffix *-ment*

This suffix forms nouns from the corresponding verbs, e.g. 'measure*ment*' (l.16) from the verb (*to*) *measure*.

EXERCISE

By adding *-ment*, form nouns from the following verbs: equip; move; adjust; establish; attach; improve; state.

3 The suffix *-ity*

This suffix forms nouns from the corresponding adjectives, e.g. 'activ*ity*' (l. 28) from the adjective *active*; 'probab*ility*' (l. 26) from *probable*, and 'simpl*icity*' (l. 23) from *simple*.

EXERCISE (a)

Form nouns from the following: alkaline; relative; potential; complex; equal; reliable; acid.

NOTE: the adjective *able* becomes 'ab*ility*'.

(b)

Applying the principle given in the Note above, make adjectives corresponding to the following nouns: availability; adaptability; stability; responsibility.

4 The prefix–suffix *-en*

This is used either as a prefix to adjectives (or occasionally nouns) to form a verb (e.g. '*en*able' (l. 32), '*en*large', etc.) or more commonly as a suffix, e.g. 'wid*en*' (from *wide*).

EXERCISE

By adding *-en*, form verbs from the following:

length; strength; tight; weak; loose; short; deep; height.

REVISION EXERCISE

Complete the following by choosing appropriate words from **1**, **2**, **3** and **4** above:

The main a...ity of the scientist is the i...ion and e...ion of the world around us. To en... him to do this he uses many different kinds of e...ment, and in order to make them more a...able to his purposes he frequently makes a...ments to them which lead to their i...ment. For example, he may s...en a part which is too weak, l...en one which is too short and t...en something which is too loose, and thus causes too much m...ment, so that the instrument does not have the necessary s...ity. So even the most specialized scientist needs to be an engineer, sometimes!

Structure Study

THE PASSIVE

The main structure used in this passage is the Passive of the tense used in Unit 1, i.e. the Simple Present. We use the Passive when we have little interest in or knowledge of, the doer of the action but are more interested in what happens to, or is done to, the person or thing thus affected.

You probably remember that the Passive is formed by the appropriate tense of the verb *to be* plus the Past Participle, e.g.

	THE ACTIVE	BECOMES IN THE PASSIVE		
People *use*	statistics	Statistics	*is*	USED
	mathematics	Mathematics		
	imagination	Imagination		
	decimals	Decimals	*are*	
	computers	Computers		
	chemicals	Chemicals		

c.f. 'Other branches of mathematics *are used* in many sciences' (ll. 24–25). The passive is used here because we are not at the moment concerned with *who* uses these branches.

Similarly: 'Attempts *are* constantly *made* to deceive the ordinary citizen' (ll. 34–35). We do not wish to specify at this point *who* makes these attempts.

EXERCISE (a)

Make the following sentences Passive, thus eliminating the unspecified doer of the action, and emphasizing the object, or the main verb:

1 People apply mathematics in many different activities. (Begin: Mathematics is ...).
2 People use the binary scale in electronic computers.
3 People form the square of a number by multiplying the number by itself. (Begin: The square of a number is ...).
4 In the binary scale, people express numbers by combinations of 0 and 1.
5 People usually use decimals rather than fractions for scientific purposes. (Begin: Decimals, rather than fractions ...).

6 People develop new products every day.

7 People call mathematics 'the language of science'.

8 People use the decimal system even in countries with non-decimalized systems of weights and measurements.

9 It is easier to perform mathematical operations with computers if we use the binary system instead of the decimal system.

10 People use electronic computers for many different purposes.

11 People often find relationships in incomplete data.

12 People make attempts to deceive the ordinary citizen.

NOTE: If the doer of the action has some importance (though less than the object), or is needed to complete the sense of the sentence, it is given, e.g. 'A knowledge of statistics is required *by every type of scientist*' (ll. 29–30).

Notice that there is a small problem of word-order in all but the most simple form of this type of sentence,

e.g. The scientific investigator applies logical and persistent thought to his problems (Active), becomes:

Logical and persistent thought is applied by the scientific investigator to his problems (Passive).

The order of words is thus: object—verb in the Passive—subject—rest of sentence.

(b) Make the following sentences Passive, mentioning the doer of the action but shifting the emphasis to the object:

1 A combination of the digits 0–9 forms integers.

2 Engineers require an advanced knowledge of algebra and geometry. (Begin: An advanced knowledge of ...)

3 Scientists, especially physicists and engineers, often use electronic computers.

4 Journalists, who seldom have a knowledge of statistics, frequently mislead the ordinary citizen.

5 Every day, applied scientists and technologists produce new drugs, fibres, chemicals and equipment. (Begin: Every day, new ...)

6 A combination of two elements forms a chemical compound.

7 The ordinary citizen often requires an elementary knowledge of statistics.

8 Economists also use mathematics.

9 Every type of scientist requires a knowledge of statistics.

10 Scientists use accurate systems of measurement.

11 Philosophers employ specialized extensions of mathematics.

12 Physicists also use probability theory.

(c) Make up sentences similar to the ones given in Exercises (a) and (b) above, using words learnt in this unit and Unit 1, and then change them from Active to Passive.

SUBSTITUTION TABLE

NOTE: Sentences must be made sensible by using an *appropriate* qualifying word from Column 3 in each case.

SIMPLE PRESENT PASSIVE

1	2	3	4	5	6	7	8	9
Logical thought Patience Accurate observation	is	always usually often sometimes never	used applied employed	by	a scientist him an engineer	in	his	work
Computers New techniques Reliable instruments	are				technologists them statisticians		their	

Make the above (a) negative, (b) interrogative.

Discussion and Criticism

1 Try to explain the nature and use of: geometry; algebra; statistics; probability theory. Give actual examples where possible.

2 In what ways are any of the branches of mathematics mentioned in the passage connected with the science you study yourself? Give examples.

3 Try to (a) state, (b) prove: (i) an algebraic formula; (ii) a geometrical theorem; (iii) a physical, chemical or biological law.

4 Comment on the graphs below, which appeared in a Government and Opposition newspaper respectively just before a General Election. They represent the Government's spending on science during its 4 years of office.

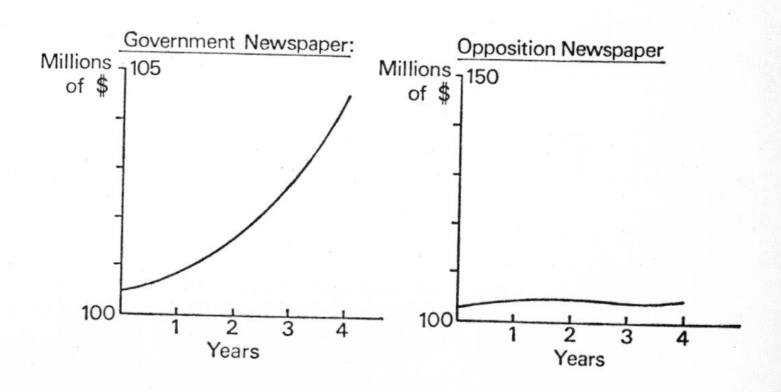

5 How can mathematics be made more interesting, especially to children?

6 Find, and comment on, some examples of statistics used misleadingly by journalists and others (there are usually plenty in newspapers and magazines).

7 What kind of people or organizations deliberately try to deceive the ordinary citizen by using statistics wrongly? Why and how?

Unit 3

SCIENTIFIC METHOD AND THE METHODS OF SCIENCE

It is sometimes said that there is no such thing as the so-called 'scientific method'; there are only the methods used in science. Nevertheless, it seems clear that there is often a special sequence of procedures which is involved in the establishment of the working principles of science. This sequence is as follows: (1) a problem is recognized, and as much information as appears to be relevant is collected; (2) a solution (i.e. a hypothesis) is proposed and the consequences arising out of this solution are deduced; (3) these deductions are tested by experiment, and as a result the hypothesis is accepted, modified or discarded.

As an illustration of this we can consider the discovery of air-pressure. Over two thousand years ago, men discovered a method of raising water from one level to another by means of the vacuum pump. When, however, this machine passed into general use in the fifteenth and sixteenth centuries, it was discovered that, no matter how perfect the pump was, it was not possible to raise water vertically more than about 35 feet. Why? Galileo, amongst others, recognized the problem, but failed to solve it.

The problem was then attacked by Torricelli. Analogizing from the recently-discovered phenomenon of water-pressure (hydrostatic pressure), he postulated that a deep 'sea of air' surrounded the earth; it was, he thought, the pressure of this sea of air which pushed on the surface of the water and caused it to rise in the vaccum tube of a pump. A hypothesis, then, was formed. The next step was to deduce the consequences of the hypothesis. Torricelli reasoned that this 'air pressure' would be unable to push a liquid heavier than water as high as 35 feet, and that a column of mercury, for example, which weighed about 14 times more than water, would rise to only a fourteenth of the height of water, i.e. approximately 2.5 feet. He then tested this deduction by means of the experiment we all know, and found that the mercury column measured the height predicted. The experiment therefore supported the hypothesis. A further inference was drawn by Pascal, who reasoned that if this 'sea of air' existed, its pressure at the bottom (i.e. sea-level) would be greater than its pressure further up, and that therefore the height of the mercury column would decrease in proportion to the height above sea-level. He then carried the mercury tube to the top of a mountain and observed that the column fell steadily as the height increased, while another mercury column at the bottom of the mountain remained steady (an example of another of the methods of science, the controlled experiment). This further proof not only established Torricelli's hypothesis more securely, but also demonstrated that, in some aspects, air behaved like water; this, of course, stimulated further enquiry.

Comprehension

1 What does the establishment of the working laws of science often involve?

2 What does a scientist collect when he tries to establish a scientific law?

3 What is the next step in the process described above?

4 What does the scientist then deduce?

5 How does he proceed to verify these deductions?

6 What does he finally do with his original hypothesis?

7 Give an approximate date for the invention of the vacuum pump.

8 Is it possible to raise water from the bottom floor of a building to the roof 50 feet above, using a vacuum pump? Why?

9 What was Torricelli's theory about the height of the water in a vacuum tube?

10 What were his deductions concerning the effect of air pressure on a column of mercury?

11 What further inference was made by Pascal?

12 Why did he use *two* mercury tubes?

13 What were the three results of Pascal's experiment?

14 What do you think happened to the mercury column when it was carried down the mountain?

Word Study

EXERCISE

Using appropriate words chosen from the reading passage, fill in the blanks in the following:

The scientist or technologist uses many m...s when he tries to s... a problem. For instance, an engineer who wants to r... a l... from one l... to another has the choice of several different p...s. One of them is to use a p... which takes the air out of the pipe or t... along which he wants the l... to flow, thus creating a v.... Air p... then pushes on the lower s... of the l... and forces it up the pipe. This method is d...d in the petrol system of a car.[1]

OPPOSITES EXERCISE (a)

From the reading passage choose words which mean the opposite of the following:

shallow (d...p); to lower (to r...e); to rise (to f...l); high (l...w); to succeed (to f...l); to refuse or reject (to a...t); imperfect (p...t); irrelevant (r...t); to pull (to p...h); depth (h...t); horizontal (v...l); to increase (to d...e); seldom (o...n).

(b) Use the words given above in sentences, using the Present Tense and Present Tense Passive.

WORD-BUILDING

The suffix *-ize*[2]

This forms verbs from nouns and adjectives, and has the meaning: to cause to be or have, or: to subject to a process of, e.g.

[1] *gasoline* system (U.S.) [2] frequently spelt -ise

analogizing (l. 20) is equivalent to: subjecting (the problem) to a process of analogy.

EXERCISE

By adding *-ize*, form verbs from the following: standard; special; local; pressure; theory; sterile; popular; familiar; neutral; optimum.

NOTE: anal*yse*, from analysis; paral*yse*, from paralysis; minim*ize*, from minimum; maxim*ize*, from maximum; and *util*ize, from use.

In technical literature this suffix is sometimes used with the names of persons or places associated with certain processes, e.g. *macadamize* (road engineering), *pozzuolize* (geology and engineering), and *pasteurize* (food technology).

Structure Study

SIMPLE PAST TENSE

The main structure used in the passage of Unit 3 is the Simple Past Tense. You will probably remember that this is the tense normally used for describing actions which happened in the past and are now finished. With regular verbs the tense is formed by adding *-ed* or *-d* (if the infinitive already ends in *-e*) to the infinitive, e.g.

'Men discover*ed* a method of raising water' (ll. 12–13)

With other subjects, the verb is still in the same form, e.g.

I You We They	discovered	a method, etc.

There are, however, a number of irregular verbs which are frequently encountered, and these have their own special forms of past tense and participle. A list of the most common is given in Appendix B, and should be revised now.

EXERCISE (a) Repeat the first exercise in the Word Study section above, putting the verbs into the Past Tense.

(b) Put the reading passage of Unit 1 (*The Scientific Attitude*) into the Past Tense.

The Negative

The Simple Past Tense, with both regular and irregular verbs forms the negative by the use of *did not* before the infinitive of the main verb; this is the same for all subjects, e.g.

Galileo I You We They	*did not* SOLVE the problem of the water-pump

The Interrogative

The Simple Past Tense forms questions by the use of *did* before the subject of the main verb, e.g.

Did [Pascal / I / you / we / they] TEST his inference by means of an experiment?

REVISION EXERCISE

Put the following sentences, which contain irregular verbs, into (i) the Simple Past Tense; (ii) Simple Past Negative; (iii) Simple Past Interrogative:

1 The liquid rises in the tube.
2 The pipes bend under the weight, and break.
3 The aircraft flies faster than sound.
4 The electric motor drives a pump.
5 The engineer takes a lot of measurements.
6 The scientist chooses between several procedures.
7 The hot-water system loses a lot of heat.
8 The lazy schoolboys make an electronic computer.
9 Later, they become famous scientists.
10 We give the results of the calculations in decimals.
11 The experiment takes a long time to carry out.
12 They draw many inferences from the hypothesis.
13 I find the ratio between the numbers.

THE PASSIVE OF THE SIMPLE PAST TENSE

This is used for the same purposes as the Simple Present Passive (see Structure Study section of Unit 2), except that it refers to the past, not the present. It is also formed in the same way, except that *is* and *are* are replaced by *was* and *were* respectively, e.g.

'It *was discovered* that it was impossible to raise water more than about 35 feet' (ll. 15–17)

The problem *was* then *attacked by* Torricelli. (l. 20)

EXERCISE

Put exercises (a) and (b) of the Structure Study section of Unit 2 into the Simple Past Passive.

SUBSTITUTION TABLE

Simple Past Tense

A Affirmatives

1	2	3	4
I	assembled	the electronic unit	last month
We	completed	the apparatus	several weeks ago
He (she)	obtained	a new measuring device	in 1964
They	built	the equipment	some time ago
The technicians	finished	the experimental model	at the end of 1963
Our research group	began work on	the prototype	a fortnight ago
An investigator	tested	a new system	at the beginning of last year

B Negatives

1	2	3	4
He (she)	did not	find the ratio between the two quantities	the last time the experiment was performed
The investigators		obtain the right figures	
The specialist		make any mistakes	
We		give the results in decimals	
The student		set up sufficient controls	
They		see the implications of the problem	
The researcher		write down all the data	

C Questions

1	2	3	4
Did	you	understand the main problem	in last week's experiments?
	he	spoil a specimen	
	the researchers	choose the most efficient procedure	
	the technician	set up the apparatus correctly	
	they	draw the right conclusions	
	our team	test the new measuring device	
	the students	make any mistakes	

D Passives

1	2	3	4
In the experiments we did last year	some new apparatus a fresh approach an interesting theory	was	developed employed tested
	some complex instruments several basic concepts many different methods	were	

Additional Exercise (for irregular verbs): Put the sentences of Table B above into the Affirmative (i.e. Simple Past Affirmative).

Discussion and Criticism

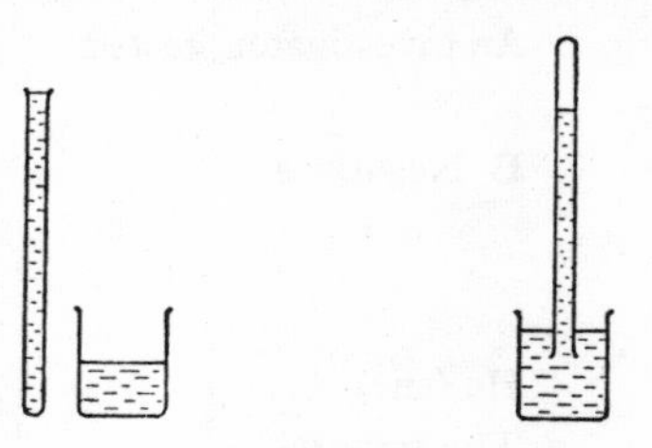

From the simple diagrams given above, describe Torricelli's famous experiment in a clear and orderly way. This should include his method of forming a vacuum.

2 Draw a simple diagram, or series of diagrams, of an experiment you know. Exchange this with another student (or group): you then describe his experiment, while he explains yours.

3 Torricelli's experiment not only provided support for his hypothesis but also involved the invention of a basic scientific instrument, the barometer. What other basic instruments do you know? How do they work, and what are some of their uses?

4 In the description of scientific method (ll. 5–10), step 2 says 'a solution is proposed'. However, in practice it is common to find that several solutions seem to be equally possible (the multiple hypothesis). How would you proceed in such a case?

5 Given an example of scientific method used in the development of the science you study yourself.

6 Do you agree that there is no *one* scientific method? Give reasons and examples.

7 What do you think is meant by 'as much information as appears to be relevant is collected' (ll. 6–7)? What was the relevant information in Torricelli's case? (Note the developments in hydrostatics.)

BIBLIOGRAPHY

WILSON *Introduction to Scientific Research*, McGraw-Hill.

BEVERIDGE *The Art of Scientific Investigation*, Heinemann (U.K.); Vintage Books (U.S.)

CONANT *Science and Common Sense*, Yale Paperbacks.
Understanding Science, Mentor.

Unit 4

PURE AND APPLIED SCIENCE

As students of science you are probably sometimes puzzled by the terms 'pure' and 'applied' science. Are these two totally different activities, having little or no interconnection, as is often implied ? Let us begin by examining what is done by each.

Pure science is primarily concerned with the development of theories (or, as they are frequently called, models) establishing relationships between the phenomena of the universe. When they are sufficiently validated, these theories (hypotheses, models) become the working laws or principles of science. In carrying out this work, the pure scientist usually disregards its application to practical affairs, confining his attention to explanations of how and why events occur. Hence, in physics, the equations describing the behaviour of fundamental particles, or in biology, the establishment of the life cycle of a particular species of insect living in a Polar environment, are said to be examples of pure science (basic research), having no apparent connection (for the moment) with technology, i.e. applied science.

Applied science, on the other hand, is directly concerned with the application of the working laws of pure science to the practical affairs of life, and to increasing man's control over his environment, thus leading to the development of new techniques, processes and machines. Such activities as investigating the strength and uses of materials, extending the findings of pure mathematics to improve the sampling procedures used in agriculture or the social sciences, and developing the potentialities of atomic energy, are all examples of the work of the applied scientist or technologist.

It is evident that many branches of applied science are practical extensions of purely theoretical or experimental work. Thus the study of radioactivity began as a piece of pure research, but its results are now applied in a great number of different ways—in cancer treatment in medicine, the development of fertilizers in agriculture, the study of metal-fatigue in engineering, in methods of estimating the ages of objects in anthropology and geology, etc. Conversely, work in applied science and technology frequently acts as a direct stimulus to the development of pure science. Such an interaction occurs, for example, when the technologist, in applying a particular concept of pure science to a practical problem, reveals a gap or limitation in the theoretical model, thus pointing the way for further basic research. Often a further interaction occurs, since the pure scientist is unable to undertake this further research until another technologist provides him with more highly-developed instruments.

It seems, then, that these two branches of science are mutually dependent and interacting, and that the so-called division between the pure scientist and the applied scientist is more apparent than real.

Comprehension

1 What is often implied by the terms 'pure' and 'applied' science?
2 What is the aim (object) of pure scientific investigation?
3 Name some examples of basic research.
4 How are the working laws of science established?
5 What is the work of an applied scientist?
6 Name some examples of applied science.
7 Name some applications of radioactivity.
8 Name some examples of the interaction of pure and applied science.
9 Give two other words meaning the same thing as hypothesis.

Word Study

EXERCISE

Complete the following sentences, choosing *one* of the four expressions in the brackets:

1 The results of research into radioactivity are applied in (electronic computers; sampling procedures; cancer treatment; pure science).

2 Many branches of applied research developed out of (the work of technologists; pieces of basic research; equations describing the behaviour of fundamental particles; new processes).

3 Pure science relates to (more highly-developed instruments; sampling procedures; solving practical problems; developing theories which explain the relationships between phenomena).

4 New kinds (types) of instruments are frequently essential for (developing basic research; improving fertilizers in agriculture; describing the life cycles of insects; finding the cube root of fractions).

5 Investigating the strength and uses of materials is an example of (the principles of pure science; technology; the interaction of basic and applied research; a theoretical model).

NOUNS AND THEIR ASSOCIATED VERBS

To use a language properly, it is important to know not only the names of things (nouns) but also the names for the actions that are associated with them (verbs): the actions are as important as the objects. Here is a list of the verbs connected with some important nouns appearing in this unit and also Units 1 and 2:

to obtain [evidence / knowledge / information / results]

to [invent / design / develop / modify] [a machine / an instrument / a process / a technique]

to	design, plan make, perform, conduct, carry out control time repeat	an experiment

to	develop suggest prove, validate disprove modify discard support put forward test	a theory a hypothesis

EXERCISE (a) Complete the following sentences with suitable verbs from the above tables:

1 A scientist must ... adequate evidence to ... a theory.

2 We must ... many experiments in order to ... a new process.

3 If an experiment is not successful, we must ... it.

4 An experiment must be carefully ...ed if we want it to ... a theory properly.

5 Technologists ... new machines to increase production.

6 If a series of carefully ...ed experiments dis... a hypothesis, we should ... it.

7 Engineers ... experiments to ... information about the strength of materials.

8 When new instruments are ...ed, the scientist is able to ... further experiments which frequently have the result of ...ing or ...ing well-established theories.

(b) What verbs are associated with the following nouns? (They all appear either in this unit or Units 1 and 2):

a próblem; observations; research; a statement; relationships; mathematical operations.

WORD-BUILDING

1 The suffix *-al*. This forms adjectives from the corresponding nouns, e.g. 'practic*al*' (l. 11) from *practice*, 'theoretic*al*' (l. 29) from *theory*. Adjectives from the names of sciences ending in *-ics* also take this suffix, e.g. mathematics—mathematic*al*.

NOTE: theory—theor*etical*; geometry—geometr*ical*; hypothesis—hypoth*etical*; techni*que*—techni*cal*; machine—mech*anical*; centre—cent*ral*; air—a*erial*; cycle—cycl*ical*.

EXERCISE Form further adjectives from the following:

addition; condition; experiment; nature; neuter; operation; section; region; analysis; matter.

2 The prefix *inter-*. This is added to verbs and derivatives to give the extra meaning of: between, among, one with the other, e.g. *interconnection* (l. 3), *interaction* (l. 37).

EXERCISE (a) Form adjectives from the following:
dependent; related; national.

(b) Form verbs from the following, using the prefix *inter-* in all cases:
act; breed; change; connect.

Structure Study

THE *-ing* FORM (I)

The main structure used in the passage of Unit 4 is the *-ing* form of the verb. This is frequently used by scientific writers because of its conciseness and flexibility, and is employed in a number of different ways. Note the following examples:

(i) 'Are these two totally different types of activity, *having* ... no interconnection?' (l. 2)
'The equations *describing* the behaviour of fundamental particles.' (l. 12)
In both these cases the *-ing* form takes the place of a longer phrase with *which, who* or *that*. Thus in the first example *having* is equivalent to *which have*; in the second, *describing—which describe.*[1]

(ii) 'Such activities as *investigating* the strength ... of materials, *extending* the findings of pure mathematics ... and *developing* the potentialities of atomic energy ...' (ll. 22–26)
Here, the *-ing* form takes the place of the derived noun: *investigating* = *the investigation of*, *extending* = *the extension of*, etc.

(iii) 'These theories ... become the *working* laws of science.' (l. 8).
'These two branches of science are mutually dependent and *interacting*.' (l. 45–46)
In the above examples, the *-ing* structure is used as an adjective describing (which describes) the noun it is associated with.

(iv) 'Let us begin by *examining* what is done by each.' (l. 4).
'(Radioactivity is applied) in methods of *estimating* the ages of objects.' (l. 34)
Note that in these cases the *-ing* form follows a word like *by, of, with, from, in,* etc. (prepositions). Many nouns, verbs and adjectives are associated with prepositions that complete their meaning, and any verb following (which follows) these prepositions takes the *-ing* form.

(v) 'The technologist, in *applying* a particular concept of pure science ... reveals a ... limitation in the theoretical model.' (ll. 38–39).
Here the *-ing* form is used, in association with a preposition, in place of a longer phrase with a noun or verb. Thus in the example given above, *in applying* is equivalent to: during the process of the application of ...

[1]A slightly different case occurs in l. 11, where the verb '*confining*' refers to 'the pure scientist', and is equivalent to the phrase '*and confines*'.

NOTE: The *-ing* form is also used in two additional cases which are not illustrated in passage 4. These are:

(vi) As part of the Continuous (Progressive) Tenses,

(vii) After certain verbs, such as *avoid*.

These uses are illustrated in Units 6 and 7 respectively.

Focus your attention for the moment on the first two uses demonstrated above:

1 As a replacement of (replacing) a phrase with *who, which* or *that*

EXERCISE (a) Find further examples of this use in the passage.

(b) In the following sentences, replace the phrases in italics with the appropriate *-ing* form:

1 A person *who does* research in chemistry is called a research chemist.

2 The research scientist often comes across problems *that require* new types of instrument for their solution.

3 New types of instrument frequently lead to discoveries *which modify* the basic principles of science.

4 Scientists sometimes develop theories *which affect* other human activities such as morals or religion. (Do you agree that morals and religion are 'activities'?).

5 Technologists develop new techniques *which increase* man's control over his environment.

6 Theories *that describe* the nature of the universe are constantly revised by scientists.

7 The force *that holds* the solar system together is gravitation.

8 The total amount of chemical reactions *that take* place in a living organism is its metabolism.

9 Viruses are entities *that occupy* a position between living and non-living matter.

10 Scale models *that reproduce* the behaviour of flowing water are used in hydraulics research.

11 Some rockets use liquid fuels *that consist* of oxygen and kerosene.

12 Newton described the laws *that govern* the motion of falling bodies.

2 Replacing a noun

EXERCISE (a) Find further examples of this use in this passage and also in the passage of Unit 1 (*The Scientific Attitude*).

(b) In the following passage replace the word or phrase in italics by the appropriate *-ing* structure:
The work of the technologist is *the application of* the theories of

the research scientist, *the development of* new processes, *the invention of* new machines and *the extension of* the uses of techniques *which exist* already. It is often difficult to separate his work from some of the activities *which belong* to the pure scientist, such as *the design of experiments* and *the elaboration of* hypotheses.

Discussion and Criticism

1 How are the following sciences applied for technological purposes: geology; meteorology; chemistry; psychology? Give details.

2 Do you agree that many pieces of applied research began as pure science? (ll. 28–29) Give examples.

3 Name some materials used in engineering. Why is it important to test their strength?

4 Give any details you know about an inventor and his work, and if possible, about its connection with basic research.

5 Do you know any examples of an advance in the field of pure science which was dependent on the development of new instruments? (ll. 41–44).

6 Give examples of how the following are applied in the discipline you study yourself: radioactivity; statistics; optics; electricity; magnetism; psychology.

7 Do you agree with the conclusions of the last paragraph? (ll. 45–48). Give reasons for your answer.

8 Radio, television (TV) and films often give a favourable picture of the pure scientist, and an unfavourable one of the applied scientist (excluding doctors). Is this true in your own country? Why? Give your own opinion in the matter.

9 Give examples of man's increasing control over his environment.

Unit 5

(Revision of material appearing in Units 1–4)

'DIRECTED' RESEARCH?

A recent phenomenon in present-day science and technology is the increasing trend towards 'directed' or 'programmed' research; i.e. research whose scope and objectives are predetermined by private or government organizations rather than researchers themselves. Any scientist working for such organizations and investigating in a given field therefore tends to do so in accordance with a plan or programme designed beforehand.

At the beginning of the century, however, the situation was quite different. At that time there were no industrial research organizations in the modern sense: the laboratory unit consisted of a few scientists at the most, assisted by one or two technicians, often working with inadequate equipment in unsuitable rooms. Nevertheless, the scientist was free to choose any subject for investigation he liked, since there was no predetermined programme to which he had to conform.

As the century developed, the increasing magnitude and complexity of the problems to be solved and the growing interconnection of different disciplines made it impossible, in many cases, for the individual scientist to deal with the huge mass of new data, techniques and equipment that were required for carrying out research accurately and efficiently. The increasing scale and scope of the experiments needed to test new hypotheses and develop new techniques and industrial processes led to the setting up of research groups or teams using highly-complicated equipment in elaborately-designed laboratories. Owing to the large sums of money involved, it was then felt essential to direct these human and material resources into specific channels with clearly-defined objectives. In this way it was considered that the quickest and most practical results could be obtained. This, then, was programmed (programmatic) research.

One of the effects of this organized and standardized investigation is to cause the scientist to become increasingly involved in applied research (development), especially in the branches of science which seem most likely to have industrial applications. Since private industry and even government departments tend to concentrate on immediate results and show comparatively little interest in long-range investigations, there is a steady shift of scientists from the pure to the applied field, where there are more jobs available, frequently more highly-paid and with better technical facilities than jobs connected with pure research in a university.

Owing to the interdependence between pure and applied science (see Unit 4), it is easy to see that this system, if extended too far, carries considerable dangers for the future of science—and not only pure science, but applied science as well.

Comprehension

1 What is programmed research?

2 What differences in working conditions are there between

the present-day scientist and scientists working at the beginning of the century?

3 Describe laboratory conditions at the beginning of the century.

4 What were the origins of programmed research?

5 Why is it difficult nowadays for the individual scientist to make significant contributions to science?

6 Mention one of the effects of organized research on the attitudes of scientists.

7 What is a common attitude of private industry and government departments towards scientific investigation?

8 What part does money play in the situation discussed in the passage?

9 How is the situation likely to affect the future of science?

10 Give another word meaning the same as 'applied science'.

11 Give two other words for 'directed' research.

Word Study Revision

EXERCISE (a) The reading passage contains numerous examples of suffixes and prefixes used in Units 1–4. Pick these out, and give the meaning of the prefix or suffix in each case.

(b) Give the opposites of: suitable; likely; frequent; limited; essential; able; efficient.

(c) In the following sentences, use a verb with *en* as a prefix or suffix to replace the expression in italics:

1 They *increase the length of* the pipe.

2 We *made* the road *wider*.

3 The engineers *increase the strength of* the bridge.

4 That government department plans to *make* its laboratories *larger*.

5 The tube was *made shorter*.

6 The high temperature had the effect of *making* the metal *softer*.

7 The softening of the metal had the effect of *making* the whole structure *weaker*.

(d) Add the appropriate suffixes to form the names of specialists in the following scientific disciplines: archaeology; obstetrics; ecology; agronomy; economics; physics; statistics.

(e) Using nouns formed from verbs given in the exercises in Unit 1, complete the following:

1 The po...ion of England is about 50,000,000.

2 Governments talk about a re...ion of nuclear armaments.
3 There is a close re...ion between pure and applied science.
4 The use of radio was responsible for a great increase in the speed of tr...ion of messages.
5 What are the main op...ions of arithmetic?
6 Many new devices were used in the co...ion of the latest type of computer.
7 Atmospheric pressure varies considerably, but these fl...ions can be recorded by means of a barograph.

(f) Using nouns formed in the appropriate exercise of Unit 2, complete the following:

1 The accurate me...ment of quantities is very important in science.
2 A good scientist is highly critical of his own st...ments.
3 Scientific instruments and machines frequently need ad...ment before they are used.
4 The de...ment of scientific e...ment is a specialized process.
5 Experimental methods often lead to the es...ment of working principles or laws.
6 One of the aims of programmatic research is the im...ment of industrial techniques.

(g) Give the opposites of: tight; to raise; deep; often; horizontal; regular; to increase.

Structure Study Revision

1 Simple Present Tense

EXERCISE (a) Make the following sentences interrogative:

1 She wants to know the answer to the problem.
2 He carries out experiments.
3 Some scientists use complex procedures.
4 He tests his theory very carefully.
5 Good laboratory conditions act as a stimulus to research.
6 The scientist applies persistent and logical thought to his problems.
7 The experiments reveal a limitation in the theoretical model.

(b) From the above sentences, choose appropriate ones only (i.e. that make sense) to put in the negative.

2 Simple Present Passive

EXERCISE (a) Put the following sentences into the passive (decide whether an agent with *by* is necessary or not):

1 People use mathematics in all branches of science.

2 People apply scientific methods in many everyday activities.
3 People obtain a great deal of useful knowledge from the study of nature.
4 People usually use the decimal system for scientific purposes.
5 People control experiments to obtain accurate results.
6 People obtain accurate results from controlled experiments.
7 Different kinds of people often make attempts to deceive the ordinary citizen.

(b) Put the following sentences into the passive (decide whether an agent with *by* is necessary or not):

1 Government departments apply programmed research on an increasing scale nowadays.
2 Specialized technicians develop modern scientific instruments.
3 The work of the technologist frequently helps the basic scientist.
4 Nowadays social scientists investigate an increasingly wide range of problems.
5 An electric pump raises the water.
6 Geologists use radioactivity as a means of dating rocks.
7 Scientists require very strong evidence before they accept a theory.

3 Simple Past Tense

EXERCISE (a) Put the following into the Simple Past Tense:

1 The bridge bends under its own weight.
2 The electric motor drives the pump.
3 The scientist chooses between several possible solutions.
4 The Torricelli experiment becomes famous.
5 The engineers find a new method of testing metal-fatigue.
6 The water in the pump rises.
7 The pressure falls slightly.

(b) Repeat the following paragraph, putting all the verbs into the Simple Past Tense:

The geochemist goes to sea in a ship equipped with special pipes. Technicians then push these pipes through thousands of feet of water until they strike the bottom (bed) of the ocean. Then they drive the pipes into the sea-bottom, and when they bring them up again they are full of mud. The geochemist takes it to his laboratory and examines it carefully. This mud gives him evidence about the constitution of the rocks of the earth.

4 Simple Past Passive

EXERCISE Put the sentences of **2**, Exercises (a) and (b) above into the Simple Past Passive.

5 The *-ing* form of the verb

EXERCISE (a) The reading passage gives various examples of the *-ing* form. Pick these out, and replace them, where possible, by another structure having the same meaning.

(b) Replace the phrases in italics by an *-ing* form:

1 Air *that pushes* on the surface of the water causes it to rise in a vacuum pump.

2 Liquids *which weigh* more than water rise less in a vacuum tube.

3 The pressure *that exists* at the bottom of the ocean is greater than that on the surface.

4 Experiments *which proved* the effects of air pressure were conducted by Torricelli and Pascal.

5 Numbers *that consist* of digits are called integers.

6 Statistics is a discipline *which affects* all the other sciences.

7 The technologist is concerned with *the development of* new processes and techniques.

Discussion and Criticism

1 Give examples of programmed research in any field.

2 Give examples of types of research which it is difficult for a single scientist, working alone, to carry out.

3 Describe any cases you know of an individual scientist contributing to the advance of science.

4 Why does private industry want immediate results for its research?

5 Explain the last paragraph of the passage. Why may programmed research become a danger to the future of applied research? What is your own opinion?

6 To what extent is research directed or programmed in your country? Give details.

7 Do you think any research should be directed? If so, what kinds, and to what degree? Give good reasons for your answers.

8 What are the arguments for and against allowing scientists complete freedom to do the research they want to do, rather than what the Government, or some other outside person or organization, consider to be important?

BIBLIOGRAPHY GOLDSMITH & MACKAY (eds.): *The Science of Science*, Penguin Books

Science Policies and National Governments, Organization for Economic Co-operation and Development (OECD)

Unit 6

SCIENCE AND INTERNATIONAL CO-OPERATION

One of the most striking characteristics of modern science has been the increasing trend towards closer co-operation between scientists and scientific institutions all over the world.

What have been the reasons for this? One of the factors has already been discussed in Unit 5, i.e. the growing complexity and widening scope of present-day research, which has resulted in the creation of large organizations employing great numbers of scientists and technologists in programmes of directed research. This has inevitably led to the extension of many items of research beyond national boundaries.

The most important factor, however, has been the magnitude of the problems to be solved. In fact, it is becoming more and more evident that many of the problems affecting the world today cannot be solved except by the pooling of scientific effort and material resources on a world-wide scale. The exploration of space, world finance and the development of new sources of power, such as atomic energy—these are examples of areas of scientific research which are so costly and complicated that no single country or organization, working by itself, can hope to tackle them efficiently.

A third powerful reason has been the increasing political and economic interdependence of nations, both rich and poor. This has had a direct effect on large areas of scientific and technological investigation, such as those connected with armaments, communications, health, agriculture, economic planning and sociological research.

As a result of the conditions outlined above, international co-operation has been greatly intensified during the last 20 years, largely owing to the initiative of the United Nations Organization (U.N.O.) and its specialized agencies, in particular the United Nations Educational, Scientific and Cultural Organization (UNESCO). Thus the most urgent problem for many parts of the world, i.e. food production, is being dealt with by the Food and Agriculture Organization (F.A.O.). The World Health Organization (W.H.O.), another U.N. agency, not only co-ordinates many research projects on medicine all over the world, but supplies advice and aid in the control of diseases in underdeveloped areas. Technical and economic assistance is provided by other U.N. bodies such as the Economic and Social Council (ECOSOC) or the Economic Commission for Latin America (ECLA) and similar agencies for other regions of the world.

Apart from the international agencies controlled by the U.N., many scientific and technological organizations, both governmental and privately owned, are pooling their resources and incorporating themselves into supra-national bodies: a good example is the Organization for Economic Co-operation and Development, with over 20 member-countries throughout the world. Universities, too, are tending to develop joint research

50 projects with their counterparts in other parts of the world, and finally, many scientific disciplines have had, for a long time past, their own international unions and associations whose main functions are the dissemination of information, the co-ordination of research and the standardization of measurements and 55 nomenclature.

Science, then, seems to be playing a major role in the creation of the 'One World' of the statesmen's dreams.

Comprehension

1 What has been one of the increasing trends in modern science?

2 What are the three main reasons for the increase in international co-operation in science and technology?

3 Name some fields in which this co-operation has been carried on.

4 What part has the U.N. played in this?

5 Name some U.N. agencies, and say what work each has carried out.

6 Apart from the U.N. agencies, name 3 other types of organization (body, entity) which are concerned with international co-operation.

7 What type of organization is concerned with the standardization of scientific nomenclature?

8 What statesmen's dream is science making into a reality?

Word Study

COMPOUND NOUNS AND NOUN PHRASES

Compound nouns and noun phrases, i.e. nouns formed by two or three nouns standing together, of which the first one or two act as adjectives for the last, appear several times in this passage, e.g. *food production* (l. 33), *World Health Organization* (l. 35). This device is frequently used in scientific English for the usual reason, i.e. conciseness: the compound noun is shorter than the corresponding phrase. Thus *food production* is equivalent to: *the* production *of* food; *World Health Organization* is short for: The organization *concerned with the* health *of the* world. Similarly, the phrase *the control of diseases* (l. 37) could be turned into a compound noun: diseas*e* control (note that the adjectival noun usually does *not* take a plural form). Scientific writers are, in fact, inclined to use this device to excess, as in the phrase *X-ray diffraction crystal structure analysis*, i.e. the analysis of the structure of crystals by means of the diffraction of X-rays.

EXERCISE (a) Pick out additional compound nouns in the reading passage.

(b) Take words from List A and add them to the appropriate words in List B to form compound nouns (e.g. cancer research):

A		B	
cancer	mountain	column	theory
health	mercury	pump	organization
research	life	fatigue	research
probability	water	study	bottom
vacuum	metal	level	pressure
sea	air	top	cycle

(c) Form compound nouns from each of the following phrases. Remember to be careful about plurals.

1 A theory about the waves of earthquakes.
2 A study concerned with the distribution of the population.
3 A textbook about the analysis of vectors.
4 Chemistry of the nucleus of cells.
5 Techniques applicable to the breeding of plants.
6 Measurements of the transfer of heat.
7 Interpretation of photographs taken by X-rays.
8 An indicator for measuring the speed of the air.
9 Devices for the control of the flow of heat.
10 The production of machinery for the farm.

WORD-BUILDING

1 The suffix *-ify*. This forms verbs from the corresponding adjectives, e.g. *intensify* (l. 28), meaning: to make intense.

EXERCISE

Form verbs from the following adjectives:

intense; pure; simple; rare; liquid; solid; united. (There is a change of spelling for *liquid* and *united*.)

NOTE: This suffix can also be added to nouns, e.g. *exEmplify*, meaning: to form an example; *typify*: to form a type of something; and the geological and sociological term *stratify*: to form strata (layers).

2 The suffix *-ly*. This extremely common suffix forms adverbs from the corresponding adjectives, e.g. *inevitably* (l. 91), from the adjective *inevitable*, *greatly*, from *great* (l. 28), etc.

EXERCISE

Pick out further examples of adverbs ending in *-ly* appearing in this and previous units, and identify the adjectives from which they have been formed.

Structure Study

The two main structures introduced in the reading passage are the Present Perfect Tense and the Present Continuous Tense.

PRESENT PERFECT TENSE

This structure is formed by *have* or *has* with the Past Participle of the main verb, which in regular verbs is made by adding *-ed* to the infinitive e.g. 'present-day research ... *has resulted* in the creation of large organizations' (ll. 6–7).

The Past Participles of the most common irregular verbs are given in Appendix B and should be thoroughly revised *now.*

The two main functions of this tense are (i) to indicate an action happening between some indefinite time in the past and the present, e.g. in the example quoted above we are not told when exactly the organizations were created,

(ii) to indicate an action which began in the past and continues (or its effects continue) into the present, e.g. 'A third ... reason *has been* the increasing ... interdependence of nations '(l. 21). The interdependence started in the past but continues into the present.

EXERCISE (a) In the following sentences, put the verbs in brackets into the Present Perfect Tense:

1 Medicine (make) great progress in the last twenty years.

2 Scientists (study) the universe for many centuries.

3 The World Health Organization (supply) advice and aid to several countries.

4 Since it was founded in 1943, F.A.O (carry out) many projects designed to increase food production throughout the world.

5 The increasing interdependence of nations (lead) to international co-operation on a much bigger scale than hitherto.

6 Programmatic research (have) some unexpected side-effects: one of them is that scientists (become) increasingly involved in applied research, and (tend) to neglect basic research.

7 Money is a factor which (play) an important part in establishing the pattern of modern scientific investigation.

8 During the past few years, several countries (pool) their resources in order to carry out certain types of scientific investigation more efficiently.

9 Recently, more than forty countries (co-operate) in the projects connected with the International Geophysical Year (I.G.Y.).

10 Agricultural and engineering experts (make) great advances in utilizing desert areas for crop production during the last few decades.

The Negative The negative of the Present Perfect Tense is formed by putting *not* between *have* (*has*) and the Participle, e.g. Up to the present, medical researchers *have not solved* many of the problems connected with virus diseases.

The Interrogative The Present Perfect Tense forms questions by putting *have* or *has* at the beginning of the sentence, e.g. *Have* medical researchers *solved* many of the problems . . . ?

EXERCISE (b) Put the sentences of Exercise (a) above into the interrogative.

(c) Make up sentences of your own in the Present Perfect *negative*, using words chosen from the passage or Exercise (a) above.

PRESENT CONTINUOUS TENSE

This structure is formed by the present tense of *to be* (I *am*; he, she, it *is*; you, we, they *are*) plus the *-ing* form of the verb, e.g. 'Many scientific organizations *are pooling* their resources' (l. 45)

The main functions of this tense are to express actions happening at the moment of speaking or writing, and to emphasize the continuous nature of actions happening in the present. For these reasons, it is generally associated with expressions such as *at present*, *now*, *nowadays*, etc.

EXERCISE (a)

Fill in the blanks in the following sentences with the appropriate verbs from this list:

performing; developing; taking; planning; obtaining; dealing; becoming; conducting; drawing; designing.

1 The research team is ... its work for next year.
2 The engineers are ... a new type of fuel, and they are ... a lot of help from the chemists.
3 It is ... increasingly evident that many scientific problems depend upon international co-operation for their solution.
4 We are ... several new processes in our laboratory.
5 The physicist is ... further inferences from the results of his latest set of experiments.
6 Economists throughout the world are ... with the problems related with the production and distribution of goods.
7 The statistician is ... a square root by using a slide-rule.
8 The doctor is ... a delicate operation.
9 Scientists are ... more and more information about man and the universe.
10 He is ... a controlled experiment.

The Negative

The negative of this tense is formed by putting *not* between the *to be* form and the verb ending in *-ing*, e.g. They *are not working* on a problem at present.

The Interrogative

The Present Continuous Tense forms questions by putting the appropriate form of *to be* at the beginning of the sentence, e.g. *Are they working* on a problem at present?

EXERCISE (b)

Put the sentences of Exercise (a) above into the negative, then into the interrogative.

THE PASSIVE OF THE PRESENT PERFECT AND PRESENT CONTINUOUS TENSES

1 The Passive of the Present Perfect is formed by putting *been* between *have* or *has* and the Past Participle, e.g. 'International co-operation *has been ... intensified*' . . . (ll. 27–28).

2 The Passive of the Present Continuous is formed by the present tense of *to be* (*am*, *is are*) followed by *being* and the past participle of the verb, e.g. 'Food production *is being dealt* with by F.A.O.' (ll. 33–34)

SUBSTITUTION TABLES

I. Present Perfect Tense

A (Affirmatives)

1	2	3	4
The research team The scientist He (she)	has	produced evidence about made experiments on put forward a theory about	a new elementary particle corrosion in metals the structure of viruses
Our students We I	have	revised a number of concepts about suggested a hypothesis about given a description of	the effects of ultra-high speeds on human beings the new nucleic acid the origins of earthquakes

B (Questions)

1	2	3	4
Has	he (she) the group	found the answer to the problem finished building the test model completed a statistical analysis of the data	yet?
Have	you they	succeeded in determining how the reaction took place made a clear-cut decision about the future of the project	

C (Passives)

1	2	3	4	5	6	7
A new process This electronic equipment Improved apparatus	has	(recently)	been	designed developed invented	for	speeding up production reducing accident risks obtaining low temperatures
More instruments Special techniques New and more exact methods	have					

II. Present Continuous Tense

1	2	3	4	5
At present	this organization		conducting tests on	
	that department	is	concentrating research on	new materials
At the moment	these agencies	are	studying	plant mutants
	those specialists		collecting data about	food storage
				disease control

Past Continuous: Practice in this tense may be obtained by changing the time phrase in Col. 1 to read *During the whole of last year* and substituting *was* and *were* in Col. 3.

Discussion and Criticism

1 What is meant by the 'urgent problem ... (of) food protion'? (ll. 32–33) Why is it *urgent* and how do you think it can be tackled? Give as many details as you can.

2 Apart from the items mentioned in ll. 15–17, what other problems require international co-operation for their solution? Give reasons.

3 What factors tend to prevent international co-operation in science and technology, and how can these be overcome?

4 Do you agree that the political and economic interdependence of countries is increasing? Give reasons and examples to illustrate your reply.

5 What do you know about the International Geophysical Year as an example of scientific co-operation on a worldwide scale?

6 What is meant by the 'standardization of measurements and nomenclature' (ll. 54–55)? Do you think this is necessary? Give reasons and examples.

7 Give examples of international co-operation affecting your own country.

8 What international union, association or U.N. agency is connected with the discipline you yourself are studying? What sort of work does it do?

9 What is an *underdeveloped area* (l. 38)? What are some of the reasons for underdevelopment? How do you think they can be overcome?

10 Do you agree that 'science (is) playing a major role (part) in the creation of the 'One World' of the statesmen's dreams' (ll. 56–75)? Give reasons for your reply. Do you think that 'One World' is either desirable or possible?

11 Study the bar chart given below, and deduce which region(s) can be considered as underdeveloped. Give your reasoning in full.

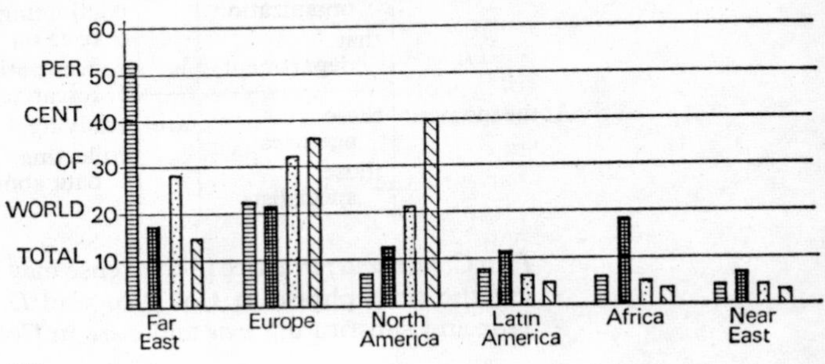

Unit 7

UNDERDEVELOPMENT AND THE SCIENCES

In this unit we are merely going to present some of the problems of underdevelopment, and then leave it to the reader to discuss how modern science can help to solve them.

To begin with, it is hardly necessary to point out that we live in a world of increasing industrialization. Whilst this process enables us to raise our standards of living at an ever-accelerating rate, it also leads to a corresponding growth of interdependence between the different regions of the world. Given (In view of) these conditions, it is easy to see that any permanent economic or political instability in one area is bound to have an increasingly serious effect upon the rest of the world. Since the main source of such instability is underdevelopment, it is clear that this now constitutes a problem of international proportions (dimensions).

What, then, is to be done? Although it is difficult to know where to begin to deal with such a large subject, the first step is perhaps to consider the main economic difficulties an underdeveloped or emerging region has to face, e.g.

(a) The economies of such countries are orientated primarily towards the production of raw materials, i.e. agricultural and mineral products; these are then exported to the industrialized countries. A number of quite common occurrences are therefore sufficient to cause immediate and serious interference with this export production: unfavourable weather conditions, plant or animal epidemics, the exhaustion of soil fertility or mineral deposits, the development of substitute products in the industrialized regions, etc. The sensitivity of the economy is greatly intensified in cases where exports are confined to only one or two products—'monocultures', as they are sometimes called.

(b) Being under-industrialized, these countries are largely dependent on imports to supply the equipment needed to produce the raw materials they export. This also applies to the manufactured goods required to provide their populations with the 'necessities of life'—a concept which is continually being enlarged through the mass media of communication such as newspapers, films, the radio and advertising. This economic structure makes it difficult for them to avoid being politically dependent on the countries which absorb their exports and provide their essential imports.

(c) Since, under modern conditions, a rapid rise in population is a phenomenon closely associated with underdevelopment (Why?), this cause alone can subject the whole (entire) economy to severe and continuous stress.

(d) Although it is obvious that industrialization is the key to development, it is usually very difficult for emerging countries to carry out plans of this nature. In the first place, to set up modern industries necessitates (requires) capital on a large scale, which only industrialized regions are able to provide; secondly, they lack the necessary trained manpower; thirdly, their industries—when established—are usually not efficient enough to compete with foreign imports, and any restriction on these imports is likely to lead to counter-action against their own exports.

From another point of view, it is necessary to bear in mind that there are invariably political, educational, social and psychological obstacles which tend to interfere seriously with any measures taken to deal with the economic difficulties outlined above. To consider only one point: it is obviously useless to devote great efforts and expense to education, technical training and planning if, for psychological reasons, the population as a whole fails to turn theory into effective action.

To conclude, it seems clear that if we are to succeed in solving the many inter-related problems of underdevelopment, only the fullest and most intelligent use of the resources of all branches of science will enable us to do so. How is this to be done? Do *you* have any suggestions to make?

Comprehension

1 Name two important consequences of industrialization.
2 Why does underdevelopment constitute a problem for the whole world?
3 What are *raw materials*?
4 What happens to the production of raw materials in an underdeveloped country?
5 Name some factors which can interfere with this production.
6 What is a 'monoculture'?
7 What kinds of products do the emerging countries import? Why?
8 Give examples of mass communication media.
9 What part do these media play in the economies of emerging countries?
10 What undesirable consequences may follow from the pattern of exports and imports of under-industrialized countries?
11 What further factor common to underdeveloped regions can put a great strain on their economies?
12 Name some factors which make it difficult for underdeveloped countries to set up industries.
13 Why is it sometimes necessary for these countries to restrict imports? Why is this a dangerous procedure?
14 Apart from economic difficulties, what other kinds of obstacle do the developing countries have to contend with?
15 What must be applied in order to solve the complex problems of underdevelopment?
16 Give other words meaning approximately the same as: whole; given; to necessitate; dimensions.

Word Study

EXERCISE (a)

Fill in the blanks using appropriate words taken from the reading passage:

The m...n method of raising st...rds of l...ng is by means of in...ation. To do this, it is ne...ry to s...up industries in order to pr...ce manufactured g...ds for the po...ation as a wh...e and eq...nt needed for the pro...tion of the raw ma...als which are ex...ted. This program(me) nec...ates large amounts of ca...l, as well as sufficient tr...ned manpower to operate these industries ef...ntly. But to ca...y out extensive edu...nal and training pro...m(me)s involves great ex...nse. Luckily, there are now several br...ches of the U.N. which can su...y the capital needed for such plans, as well as other in...nal organizations which can pr...de te...cal assistance (help). However, all this is us...ss unless the people of the cou...ies concerned tu...n these plans into ef...ive action, and this is pri...ily a psy...al problem.

(b)

As in Unit 6, some of the concepts used in the passage require strict definition. Working in groups, prepare clear definitions of: industrialization; instability (or stability); sensitivity; soil fertility; raw materials; manpower; technical training. Criticize your own definitions, and those of the other groups, mercilessly!

WORD-BUILDING

The suffix *-less*. This common suffix is added to nouns to form adjectives, having the meaning of: without, not having, e.g. *useless* (l. 57) meaning: of no use.

EXERCISE (a)

Form adjectives from the following nouns by adding *-less*. Give the meaning of each adjective you form: class; colour; effort; job; limit; motion; power; purpose; root; water; weight.

(b)

From the adjectives formed in (a), choose appropriate ones to fill in the blanks in the following:

Water is a ... liquid. A body moving in space is ... and therefore floats. A desert is almost entirely ... . At absolute temperature (A°) the molecules of a body are ... . People who are unemployed are said to be ...; because of this they have no daily objectives to attain and their lives become ...; this in turn gives rise to a social problem, since they tend to lose their feeling of stability, and become .... The prospects of improving the conditions of society by the intelligent application of science are apparently .... Do you think a ... society is possible or desirable?

PHRASAL (PREPOSITIONAL) VERBS

A common linguistic device in English as a whole is for verbs to change and extend their meanings by the addition of prepositions, or in some cases, phrases. Hence *to carry* means to support the weight of something and move it from one place to another; but *to carry* OUT (l. 45 means: to perform a job, to bring it to a conclusion. Other phrasal verbs used in the passage are *point out* (l. 4): indicate, bring to the reader's attention; *lead to* (l. 17): have as a result; *set up* (l. 45): establish; *bear in mind* (l. 153): consider or remember; and *turn into* (l. 60): transform or convert from one thing to another.

EXERCISE (a) Make up sentences containing the above phrasal verbs.

(b) Look through the Basic Dictionary for further examples of these phrasal verbs.

Structure Study

THE INFINITIVE

The main structure presented in the reading passage is the *Infinitive*. Like the *-ing* form, it is frequently used in scientific English, and for the same general reasons, i.e. because of its association with other verbal forms, and because it is a convenient substitute for longer phrases.

1 *Infinitive associated with other Verbal Forms:*

(i) After Anomalous Finites (without *to*)—

e.g. l. 3: science CAN *help* to solve them.
ll. 63–64: use of the resources ... WILL *enable* us to...

In the above examples, the anomalous finites *can* (indicating physical and mental ability) and *will* (indicating the future) are followed by the infinitive (without *to*) of the main verb. Other similar verbal forms taking the infinitive without *to* are: would, must, may, might, shall, should, and could. (For these anomalous finites and their meanings, see Unit 8.)

(ii) After certain Main Verbs (with *to*)—

e.g. l. 15: to BEGIN *to deal* with the subject.
l. 6: this ENABLES us *to raise* our standards
l. 55: obstacles which TEND *to interfere* with measures

Other common verbs taking the infinitive of an attached verb in this way are: allow, ask, attempt, decide, fail, know, how, seem, tend, try, tell, want, help, cause. The infinitive with *to* also follows certain expressions formed with the verb *to be* + adjective, e.g. be (un)able, be liable, be bound, be careful, be likely, be certain, etc.

Also, (without *to*): see, watch, hear, feel, let, make.

NOTE: Other equally common verbs are followed by the *-ing* form (see below).

(iii) After some common Impersonal Expressions (with *to*)—

e.g. l. 4: IT IS NECESSARY *to point out* ...
l. 9: IT IS EASY *to see* ...
l. 14: IT IS DIFFICULT *to know* ...

Other similar expressions (which may be negative also) are: it is (not) advisable, important, interesting, satisfactory, sufficient, usual, possible, useless. Notice that in this construction the infinitive may be part of an accusative and infinitive construction (preceded by *for*); in this case, the accusative may often be interpreted as the subject of the infinitive.

e.g. It is easy FOR industrialized countries TO raise their standard of living.

2 *Infinitive as a substitute for longer phrases:*

e.g. l. 4: *to begin* with
l. 61: *to conclude*
l. 30: the equipment needed *to produce*
l. 32: goods required *to provide*

In the above examples, the infinitive may be held to indicate purpose, and would otherwise be used with a purpose phrase, such as *in order to, so as to be able to*, etc. Alternatively, it may be viewed as a substitute for a noun phrase, e.g. *To begin* is equivalent to: As a beginning; *The equipment needed to produce* = The equipment needed for the production of, etc. In any case, the plain infinitive is shorter than the phrase for which it is a substitute.

EXERCISE (a) Read through the passage again, pointing out further examples of the different uses of the infinitive outlined above, and noting *other* uses not covered by the explanations given.

(b) From the elements given below, construct as many *sensible* sentences as you can:

	(1)		(2)		(3)	(4)
	useless		scientific disciplines		omit	industries
	difficult		emerging countries		help	international associations
IT IS	important	FOR	industrialized countries	TO	provide	an adequate standard of living for its people
	usual		a government		set up	emerging countries
	necessary		economic plans		have	to consider the psychology of the population

THE *-ing* FORM (II) Apart from the uses of this structure explained in Unit 4, there are a number of main verbs which require the *-ing* form in any verb attached to them,

e.g. ll. 36–37: it is difficult for them to AVOID *being* dependent.

Some other common verbs taking the *-ing* form are: finish, stop, risk, practise, enjoy; also verbs with an attached preposition, e.g. succeed *in* (l. 61), get used *to* (become accustomed *to*), etc. These verbs must be carefully distinguished from those which require the infinitive (see ll. (ii) above).

EXERCISE (a) Put the verbs in brackets into either the Infinitive or the *-ing* form, as necessary:

The extension of international co-operation nowadays enables the emerging countries (overcome) many obstacles to their development, and helps them (avoid) (make) serious mistakes. Any country, however, is likely (progress) more rapidly if it gets used to (make) the best use of its available resources and knows how (carry out) small-scale projects efficiently before it risks (begin) large-scale development.

(b) Put the verbs in brackets into either the Infinitive or *-ing* form, as required:

1 It is usual for scientists (use) mathematics in their work.

2 We have now finished (programme) the computer for the experiment.

3 It is difficult for many underdeveloped countries (achieve) higher standards of living because of the sensitivity of their economies.

4 We decided (conduct) trials with the equipment and eventually succeeded in (obtain) the results expected.

5 The researchers failed (locate) the failure in the control mechanism.

6 Geneticists are able (cause) certain changes in the constitution of living matter by means of X-rays.

7 This allows them (discover) further facts about the mechanisms of heredity.

8 The government asked the private company (develop) a lighter alloy for military purposes.

9 A scientist risks (make) serious mistakes if he is not careful (reason) correctly.

10 Industrialized countries should (stop) the increasing production of nuclear weapons.

11 Precision instruments allow scientists (measure) minute quantities with a high degree of accuracy.

12 The investigator decided (repeat) the experiment before (publish) the final report on his work.

SUBSTITUTION TABLES

A After certain verbs

I. The Infinitive

1	2	3	4	5	6
The specialized agency Our research group Certain international bodies Some large corporations Important scientific organizations A government department They	will is/are	begin attempt try want help likely expected	to	study the use of land analyse the effects /of radiation deal with the problem /of birth control make a survey /of natural resources	in the near future

B After certain impersonal expressions

1	2	3	4	5	6	7
It is	not difficult necessary usual important advisable	nowadays	for	government agencies research institutes large industries universities private firms	to	co-ordinate research plan a year's work in advance extend their laboratory facilities

C As substitutes for longer phrases

1	2	3	4	5	6	7	8
The	plant apparatus equipment instruments	needed required wanted installed	to	modernize expand the capacity of increase the output of maximize the efficiency of	our	factory laboratory department	will soon come into operation

D Verbs followed by the infinitive without *to*

1	2	3	4
I	watched	the student	perform the experiment again
He (she)	saw		set up a very complicated piece of apparatus
We	heard		take part in the experiment
	let		re-check the data
	made[1]		
	had[1]		

II. The *-ing* form (After certain verbs)

1	2	3
The technician	succeeded in	developing several projects every year
I	got used to	correlating data from different sources
We	became accustomed to	making controlled experiments
The students	practised	measuring very minute quantities
Our research group	finished[2]	maintaining apparatus in good condition
He (she)	stopped[2]	recording the results of the experiments

Discussion and Criticism

1 Discuss, quoting actual examples, the dangers to a developing country of: monoculture; overpopulation; lack of capital.

2 'Lack of capital is certainly a serious obstacle to development, but lack of the will or the capacity to make use of the existing resources is fatal.' Discuss and criticize.

3 Why does economic instability often lead to political instability? Give examples and explain exceptions to this general rule.

4 Discuss as clearly and concretely as possible the various ways in which the science you yourself are studying can contribute to the solution of some of the problems of development.

5 How do you think the mainly psychological problem set out in ll. 53–60 can be tackled?

6 Examine the remaining difficulties of underdevelopment and industrialization presented in the passage and suggest ways in which different sciences, working together, can help to solve them.

[1] *to make* (*have*, U.S.) *someone do something:* to compel them to do it.
[2] There is often a difference between *to finish* and *to stop*: *to finish* suggests that the process comes to a natural conclusion; *to stop* suggests that the process is interrupted.

7 Do you think there are other problems connected with the subject which have *not* been mentioned? If so, describe them clearly and give relevant examples.

BIBLIOGRAPHY

Report on the United Nations Conference on the Application of Science and Technology for the benefit of the Less Developed Areas (United Nations).

Report of the Conference on the Application of Science and Technology to the Development of Latin America (UNESCO).

Unit 8

SOURCES OF ERROR IN SCIENTIFIC INVESTIGATION

In Unit 3 we examined briefly the sequence of procedures which make up the so-called scientific method. We are now going to consider a few of the many ways in which a scientist may fall into error while following these procedures.

In formulating hypotheses, for example, a common error is the uncritical acceptance of apparently common-sense, but untested, assumptions. Thus in the field of psychology it was for many years automatically assumed that the main cause of forgetfulness as the interval of time elapsing between successive exposures to a learning stimulus. Experimentation, however, was subsequently undertaken, and several other factors, such as motivation and the strength or effectiveness of the stimulus, turned out to have an even more important bearing on the problem. A somewhat similar error arises from neglect of multiple causes. Thus two events may be found to be associated, e.g. when the incidence of a disease in a smoky industrial sector of a city is significantly higher than in the smoke-free zones. A research worker might infer that the existence of the disease is due to the smokiness of the area when in fact it might equally well be found in other reasons, such as the under-nourishment of the inhabitants or over-crowding.

Both in collecting the original evidence and in carrying out subsequent experiments, a frequent cause of error is the fact that observations are not continued for a long enough time. This may lead not only to a failure to discover positive items (e.g. Le Monnier's failure to recognize that Uranus was a new planet, not a fixed star, etc.), but may also result in important negative aspects of the investigation remaining undiscovered. In applied science, this latter error may have disastrous consequences, as in the case of the thalidomide drugs, cancer-inducing industrial chemicals, etc.

Another well-known error in experimentation is lack of adequate controls (see Unit 3, ll. 39–43). Thus a few years ago it was widely believed that a certain vaccine could prevent the common cold, since in the experiments the vaccinated subjects reported a decrease in the incidence of colds compared with the previous year. Yet later, more strictly-controlled experiments failed to support this conclusion, which could have been due to a misinterpretation of chance results. This error is often caused by a failure to test a sufficient number of subjects (inadequate sampling), a disadvantage which affects medical and psychological research in particular.

Errors in measurement, particularly where complicated instruments are used, are common: they may arise through lack of skill in the operator or may be introduced through defects in the apparatus itself. Furthermore, it should be borne in mind that apparently minor changes in laboratory conditions, such as variations in the electric current, or failure to maintain atmospheric conditions constant, may disturb the accuracy of various

50 items of equipment and hence have an adverse influence on the experiment or series of experiments as a whole. In addition, such errors tend to be cumulative.

Finally, emotion in the observer can be one of the most dangerous sources of error. This may cause the researcher to 55 over-stress or attach too much importance to irrelevant details because of their usefulness in supporting a theory to which he is personally inclined. Conversely, evidence disproving the view held may be ignored for similar reasons. Even routine matters such as the recording of data may be subject to emotional inter- 60 ference, and should be carefully checked.

To sum up (summarize), the multiple possibilities of error are present at every stage of a scientific investigation, and constant vigilance (care) and the greatest foresight must be exercised in order to minimize or eliminate them. Additional errors are, of 65 course, connected with faulty reasoning; but so widespread and serious are the consequences that may arise from this source that they deserve separate treatment in the following unit.

Comprehension

1 What is the connection between the reading passage in this unit and that of Unit 3?

2 At what stage of an investigation is the scientist most likely to commit the error of accepting untested assumptions?

3 Give an example of this type of error.

4 What type of error is similar to the above?

5 Name at least three factors that an unduly high incidence of disease in a smoky sector of a city might be due to.

6 Name two broad results that insufficient observation may lead to. Give examples of each.

7 Why was it believed that a certain vaccine could cure the common cold?

8 What is meant by *inadequate sampling*?

9 Name two causes of inaccurate measurements.

10 What other factors can affect the accuracy of instruments?

11 What is an additional danger in these so-called minor errors?

12 Name three ways in which emotion can cause scientists to make mistakes.

13 How can the possibilities of error be minimized?

14 What is the last source of error named in the passage?

Word Study

WORDS WITH DIFFERENT MEANINGS FOR THE SAME FUNCTION

A word may sometimes have more than one meaning though its function (i.e. whether it is a noun, an adjective, etc.) remains the same. An example in the reading passage is the word *constant*, which in l. 49 is equivalent to *unaltered*, whereas in l. 62 its meaning is *continuous*.

EXERCISE (a) Look up the following words in the Basic Dictionary, and note their different meanings:

to mount; power; matter; drop; plant; beam; chart; light (adj.).

(b) Complete the following sentences with words in their appropriate meanings chosen from Exercise (a) above:

1 The ... compiled by a hydrographer is different from the one a statistician uses.
2 The *Annual Bulletin* of the Department of Mines reports that there has been a ... in the production of metals during the current year.
3 The experimenters noted that the crystalline structure of the material under study was altered as the pressure ...ed.
4 In mathematics a quantity successively multiplied by itself is said to be raised to a .... Thus 4 × 4 × 4 is 4 raised to the third ....
5 Lasers are modern optical devices capable of transmitting an extremely narrow and powerful ... of light.
6 In modern engineering, the ...s used for supporting roofs are no longer made of wood, but of other materials which are stronger and ...er.
7 The laboratory specimen was ...ed on a slide for microscopic analysis.
8 Electrical ... can be generated by machines driven by the energy of moving or falling water.
9 It is a well-known fact that buildings, etc., painted in ... colours tend to reflect solar heat, while those painted in dark colours tend to absorb it.
10 The origin of life on earth is still largely a ... of speculation since no certain knowledge is available.
11 In view of the likely exhaustion of natural fuels, nuclear energy is being used for generating electricity and several countries have set up nuclear power ...s.
12 In the process of condensation, water vapour in the atmosphere cools off and condenses (i.e. it becomes liquid) around dust particles to form ...s of water which fall on the land as rain.

WORD-BUILDING

1 The suffixes *-ent* (*-ant*) and *-ence* (*-ance*). Whilst the suffix *-ent* (*-ant*) is added to verbs to form corresponding adjectives, e.g. suffici*ent* (l. 40) from the verb *to suffice*, the suffix *-ence* (*-ance*) turns both verbs and adjectives into the corresponding abstract nouns, e.g. accept*ance* (l. 6), meaning the action of accepting, exist*ence* (l. 18), meaning the quality of existing, import*ance* (l. 55), meaning quality of being important, etc.

EXERCISE (a) Using an English-English dictionary, and following the examples given in the first line of the table given below, supply the correct words in place of the question-marks (?)

persist	persistent	persistence
?	resistant	?
signify	significant	?
—	consequent	?
—	?	ignorance
interfere	—	?
—	?	convenience
depend	?	?
—	evident	?
—	present	?
diverge	?	?
differ	?	?
?	emergent	?
disturb	—	?
?	?	assistance
—	?	permanence
?	—	performance
?	convergent	?
—	resonant	?
—	?	absence

(b) From the table completed in Exercise (a) above, choose appropriate adjectives and nouns to fill in the blanks in the following sentences:

1 The em... of new nations has given rise to further international problems.

2 Economic growth leads to increased interd... between nations.

3 Agricultural researchers are now producing plants which are much more r... to disease and adverse atmospheric conditions than hitherto.

4 Economic instability may lead to political d...s.

5 If a researcher encounters dif...s between two parallel experiments, he has to find out whether they are s... or not.

6 Laboratory equipment must be designed to give a high level of p....

7 An investigator should do his best to isolate his experiments from in... due to chance or random factors.

8 It may sometimes be difficult for the investigator to determine the exact degree of relevance of the ev... he has collected.

2 The suffix *-er* (*-or*). This suffix, and its variant, forms nouns from verbs, with the general meaning of: person or thing which *-s*, e.g. work*er* (l. 5), operat*or* (l. 45), oberv*er* (l. 53) and research*er* (l. 54). Although usually referring to persons, the words so formed may also refer to things, e.g. a steriliz*er* is an apparatus for sterilizing instruments, a conduct*or* is something which conducts heat or electricity, etc.

EXERCISE (a) Add *-er* to the following verbs to form corresponding nouns:

boil; breed; count; contain; convert; enlarge; fill; produce; start; train; transmit; transform.

(b) Add *-or* to the following verbs to form corresponding nouns:

demonstrate; direct; insulate; investigate; react; indicate.

Use the words formed above in sentences of your own.

3 The prefixes *over-* and *under-*. *Over-* placed in front of a word gives the idea of excess, e.g. *over*crowding (l. 21) means excessive crowding; to *over*stress (l. 55) means to lay too much stress on. Its opposite, *under-*, gives the idea of insufficiency or inadequacy, as in *under*-nourishment (l. 20), meaning lack of sufficient nourishment.

EXERCISE (a) Put *over-* in front of the following words, and explain the expressions thus formed:

to load; to heat; acceleration; production; population; to supply; to cultivate.

(b) Put *under-* in front of the following words, and explain the expressions thus formed:

to estimate; to feed (nourish); industrialization; size; weight.

4 The suffix *-ness*. This suffix forms abstract nouns from adjectives, e.g. forgetful*ness* (l. 9), effective*ness* (l. 12), and useful*ness* (l. 56).

EXERCISE (a) Form abstract nouns from the following:

clear; cool; cold; damp; dark; direct (indirect); exact; flat; full; heavy[1]; quick; rough; shallow; sharp; short; slow; steady (unsteady); steep; thick; thin; tight; weak; light.

(b) Complete the following by choosing the appropriate words from those in **2**, **3** and **4** above.

The research ...er in the field of economics who investigates the phenomenon of under..., which leads emerging countries to rely almost exclusively on the export of raw materials for their foreign exchange, often finds that for this reason—and also to satisfy the growing demands caused by over...—both the ...ers of crops and the ...ers of animals tend to over... the land. This may lead to a steady decrease in soil fertility and is thus a cause of great ...ness in the economies of such regions. On the other hand, especially favourable climatic conditions may lead to the over... of raw materials, with a consequent drop in the export price. All these factors help the ...or to explain the ...ness and un...ness with which these countries develop, in the absence of industrialization.

Structure Study

The reading passage incorporates further examples of the *anomalous finites* first presented in Unit 7 (Structure Study

[1] Adjectives ending in *y* change to an *i* in the noun.

section, para. 1). These are words such as *may*, *might*, *can*, *could* (*be able*), *must* (*have to*), and *should* (*ought to*), which are used with the *infinitive* of a verb to modify that verb by attaching different meanings to it, as follows:

1 CAN (am, are, is ABLE)—past tense COULD (was, were ABLE). These anomalous finites convey the idea of *physical or mental ability*,

e.g. ll. 53–54. 'Emotion in the observer *can* be one of the most dangerous sources of error.'

2 COULD. Apart from forming the past tense and conditional (= would be able) of 'CAN' (see above), *could* may sometimes indicate *possibility* in the present or past, and is thus equivalent to MAY or MIGHT,

e.g. ll. 38–39: 'This conclusion *could* have been due to a misinterpretation of chance results.'

3 MAY—past tense MIGHT. In scientific English, these anomalous finites indicate *possibility*,

e.g. l. 15: 'Thus two events *may* be found to be associated ...'

4 MIGHT. In addition to forming the past tense of MAY (see above), *might* can also convey the idea of a *more remote possibility* in present and future than *may*,

e.g. ll. 18–19: 'A research worker *might* infer that the disease was due to the smokiness of the area.'

5 MUST (HAVE TO)—past tense: HAD TO. These anomalous finites indicate *necessity or compulsion*,

e.g. ll. 63–64: 'The greatest care *must* be exercised in order to eliminate errors.'

6 SHOULD (OUGHT TO). These anomalous finites (the same for all tenses) convey the idea of *moral obligation*, and are therefore less compulsive than *must*,

e.g. ll. 59–60: 'The recording of data may be subject to emotional interference, and *should* be carefully checked.'

As an extension of this idea, scientific writers frequently employ these two anomalous finites, especially SHOULD, to convey the idea of *expectation*, i.e. as a substitute for the phrase: it can be expected that,

e.g. The use of higher-octane fuels in internal-combustion engines *should* result in improved performance.

NOTE: The future tense of *can* is WILL BE ABLE TO and of *must* is WILL HAVE TO.

EXERCISE

Complete the passage below, putting the appropriate anomalous finites in the spaces according to the ideas given in brackets:

In addition to the errors mentioned in the reading passage, a scientific researcher ... (*possibility*) commit other mistakes. For example, he ... (*remote possibility*) fail to read all the relevant literature about the problem he is investigating, and so ... (*remote possibility*) miss essential pieces of evidence. Or, in designing an experiment or set of experiments, he ... (*possibility*) even incorporate the answer he expects or subconsciously desires. This ... (*physical ability*) easily happen, especially in the fields of psychology or sociology where the questionnaire method is extensively used; the investigator ... (*necessity*) be extremely careful, in framing the questions, to avoid influencing the replies of the subjects who are to answer them. To understand—and therefore avoid—these, and similar errors, the student ... (*moral obligation*) read the detailed accounts of the major scientific discoveries of the past, and wherever possible he ... (*moral obligation*) try to find out the true stories of the development of pieces of modern research by talking to the successful investigators concerned. If this procedure is followed, he ... (*expectation*) greatly increase his chances of becoming a successful worker himself.

SUBSTITUTION TABLE
Anomalous Finites

A Ability, possibility

1	2	3	4
Untested assumptions	can obviously	lead to	false generalizations
Faulty equipment	could easily	result in	invalid data
Insufficient observations	may possibly	give rise	to erroneous conclusions
Inadequate sampling	might, if uncorrected,	produce	disturbing effects in a piece of research

B Obligation, advisability, compulsion

1	2	3	4	5	6
Special care	ought, as a matter of course,	to	exercised	in	analysing course results
Critical judgement	should advisedly	be	used		developing new theories
Strict objectivity	must necessarily		applied		
Logical thought	has obviously				

Discussion and Criticism

1 Can you suggest why medical and psychological researchers (rather than, say, entomologists) are liable to fall into the error of inadequate sampling (ll. 40–43)?

2 Explain clearly what is meant by 'lack of adequate controls' (ll. 32–33).

3 Explain clearly how insufficient observation (ll. 22–28) has led to a failure to reveal weaknesses in a theory or piece of applied research. As examples of the latter, you could consider (a) the thalidomide drugs; (b) cancer-inducing industrial chemicals; (c) modern pesticides and their often disastrous side-effects; (d) aircraft or other engineering failures; (e) any other examples you know or can find out for yourself.

4 Give concrete examples of each of the errors mentioned in the reading passage or the exercise given in the Structure Study section.

5 Explain why the last line of the exercise in the Structure Study section recommends the student to try and find out the *true* stories of pieces of successful research. Why may these differ from the published accounts of the investigations?

6 Can you think of any further sources of error *not* mentioned in the unit? Try to give specific examples wherever possible.

7 The following are two different accounts of archaeological investigations at the same site (place). Study them both critically, and answer the questions attached:

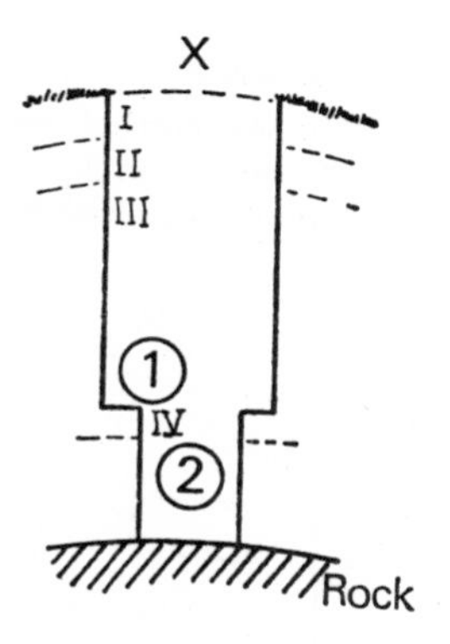

Excavation X

This was carried out by Mr Xcentric, who had been given funds by a well-known newspaper to excavate a site in the Middle East where written records indicated that a Roman settlement might have existed. He arrived on the site rather late in the season, and partly for this reason, partly because he knew the newspaper wanted quick results, he sank a shaft straight down from the top of one of the mounds (see cross-section above). He found traces of a 12th-century local settlement on top (I), then an unoccupied level (II), then abundant Roman artefacts, most of them broken, as well as layers of ashes. At ① he found a Roman coin of the 2nd century A.D. Below this he found another unoccupied level (IV), but decided he would sink a small shaft down to bed-rock to see if there were traces of further occupation. His enterprise was rewarded, and at ② he found a human bone. As he had to go on to South America almost immediately in order to do another excavation for another newspaper, he sent the bone to a local university to be dated by chemical methods. This was done, and was given an age of c. 6500 years, which corresponded fairly well with the dates of Neolithic settlements in a neighbouring country. Although rather surprised at the length of the Roman occupation implied by the depth of the layer of Roman remains, he sent his newspaper the following reconstruction:

'The site was originally settled by a Neolithic settlement, whose occupation terminated about 4500 B.C. These early inhabitants were probably primitive hunters, as no tools were found in this level. After a period of disuse, the site was occupied by the Romans, not earlier than the 2nd century A.D. This occupation continued for a considerable time, and it is presumed that, like the 'Lost Legions' of Africa, this isolated settlement kept alive the civilization of Rome for many centuries after the Roman Empire had disappeared. The many broken artefacts recovered —some of them showing traces of fire—as well as the layers of ashes indicate that the settlers suffered frequent attacks from the local inhabitants, until the end came some time before the 12th century A.D.'

Excavation Y

This was carried out by Mr Yse, an expert on Roman occupations who read Mr Xcentric's newspaper report and was also surprised at the length of the presumed Roman occupation. He

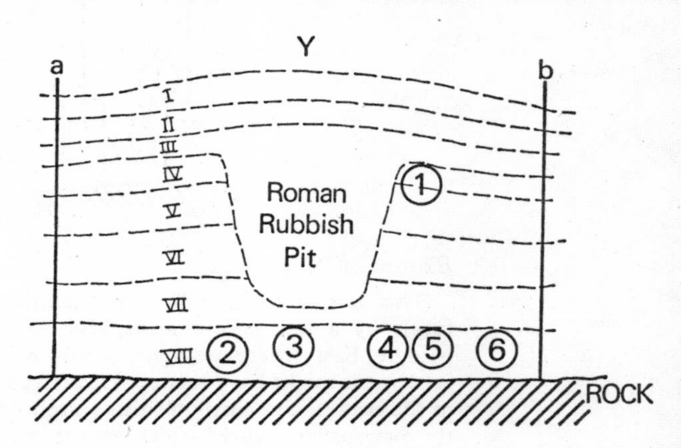

was able to persuade a benevolent Foundation to give him some money for further investigation. He thereupon chose a mound next to the one investigated by Mr Xcentric, and began to remove one by one the layers of occupation between *a* and *b*, with the results given in the cross-section:

Level I was the local 12th-century settlement; II was unoccupied; III showed traces of Roman occupation, from about the middle of the second to the beginning of the third century A.D.; IV was again unoccupied, and V showed a Scythian settlement. Throughout VI some beautifully-made stone tools, including a number of sickles for cutting crops, were found. VII was unoccupied, and at VIII human bones were found at ②, ③, ④, ⑤ and ⑥, together with a few very primitive stone tools, including a pestle and mortar. The bones were sent to Europe for dating, with the following average results for each specimen:

② 11,100 years; ③ 6,800 years; ④ 10,200 years; ⑤ 10,100 years; ⑥ 10,400 years.

At ① a broken stone sickle was found.
Mr Yse noted that the bone specimen found at ③ was directly underneath the rubbish pit and might have been affected by chemical solutions draining from the bottom of the pit.

QUESTIONS

A Explain clearly the errors committed by Mr Xcentric in his investigation. What were they due to, and how could they have been avoided?

B Write a brief account (or prepare a short lecture) on the history of the occupation of the site as Mr Yse might have described it.

C Explain the broken sickle found at ①.

BIBLIOGRAPHY

CONANT *Science and Common Sense*, Yale Paperbacks.
CANNON *The Way of an Investigator*, Norton.
HILDEBRAND *Science in the Making*, Columbia Paperbacks.
BEVERIDGE *The Art of Scientific Investigation*, Macmillan (U.K.) Vintage Books (U.S.A.).
EDGE (ed.) *Experiment—scientific case-histories*, B.B.C.
MASON *A History of the Sciences*, Collier.

Unit 9

STRAIGHT AND CROOKED THINKING

If we observe the actions of men, whether as individuals or as groups, and whether scientists or non-scientists, we find that they frequently fall into avoidable error because of a failure to reason correctly. There are many reasons for this, though only a few can be dealt with here.

The first difficulty is bound up with (related to) the use of words. It frequently happens that what one person means when he uses a certain word is different from what others mean. Consider, for example, the words *intelligence, oxygen, accurate* and *average*. In *intelligence* we face the problem that a word may not mean only one thing, but many—in this instance a very complicated set of aptitudes and abilities whose number and characteristics are not agreed upon by the specialists who study the phenomenon, and are even less understood by the layman (non-specialist). In *oxygen* we have a different problem, for although both a research chemist and a chemical manufacturer identify the word theoretically with the element O, in practice they have different concepts about it. Thus if the researcher performed a delicate experiment, using the manufacturer's *oxygen*, it might easily be a failure since the so-called O, whether used as a solid, liquid or gas, would almost certainly contain other substances. Hence another difficulty about words is that they often do not differentiate clearly enough between several varieties of the 'same' thing.

Another common error connected with words consists in confusing a word or a name with a fact. The course of scientific progress has been frequently slowed down by (1) assuming the existence of *something* to account for a certain phenomenon, (2) giving the assumed substance a name, e.g. *phlogiston, aether*, etc., and (3) implying that the phenomenon has been satisfactorily accounted for (explained).

Apart from the misuse of words, mistakes in logic can occur. Thus an example is recorded of a young sociologist, investigating literacy in a certain community, who discovered from the official records that over (more than) 50 per cent of the population were females. He subsequently found that approximately 70 per cent of the population were literate. When he had obtained this data he summed it up and drew conclusions as follow:

Most of the population are females;
Most of the population are literate;
∴ most females are literate.

This was, of course, an unreasonable inference, as the investigator himself realized as soon as he had re-examined his chain of reasoning more carefully.

Another mistake is to confuse cause and effect. This may easily occur at the beginning of an investigation, but if it remains uncorrected it can be considered as primarily a by-product of insufficient experimentation. To illustrate this, the following case can be quoted. The inhabitants of a certain community had

50 noted over the ages that whenever an individual became ill with a fever, the body parasites left him. They therefore made the correlation that the parasites kept them healthy. Later, however, properly-controlled scientific investigation showed that the reverse was true: in fact the parasites transmitted several kinds
55 of fever, and then left the sick persons when the latter's bodies became too hot to live on.

Some other factors which may influence reasoning are (a) faulty analogizing, (b) the inhibiting effect on further research of concepts which have been widely accepted as satisfactory, (c)
60 the role of authority as a bar to the re-consideration of a problem. As regards the first of these, it should be emphasized that the process of tackling one problem by analogizing from another has frequently yielded valuable results, as in the case of air-pressure (see Unit 3). On the other hand, it may lead to the adoption of a
65 totally false hypothesis, as when the idea of the atom as an infinitely small piece of solid matter was obtained by analogizing from the world of visible appearances. This erroneous viewpoint blocked progress in this field for many decades. Similarly, the comparison of the movement of light to a wave—an analogy
70 which had actually provided a satisfactory explanation of the observed phenomena during most of the nineteenth century—tended subsequently to interfere with the development of the equally valid concept of light as a stream of particles. This example also illustrates the second factor enumerated above. As
75 far as the third factor is concerned, the history of science shows many instances in which the force of authority has operated in such a manner as to build up an exceedingly powerful resistance to further investigation; in some cases centuries elapsed before this resistance was eventually broken
80 down, as happened in cosmology, for example.

Thus in addition to the chances of going astray outlined in the previous Unit, the scientific investigator shares with the ordinary citizen the possibilities of falling into errors of reasoning in the ways we have just indicated, and many others as well (in
85 addition). The more he knows of this important subject, therefore, the better equipped he will be to attain success in his work; and the straighter he thinks, the more successfully he will be able to perform his functions as a citizen.

Comprehension

1 Why do people make avoidable errors, and what sort of people make them?

2 What is the first difficulty connected with the use of words?

3 Give an illustration of this difficulty.

4 What has often been the result of the process outlined in ll. 26–31? What error does this process illustrate?

5 What evidence did the young sociologist find to support the assumptions set out in ll. 39–41?

6 What conclusions (about his inference) did he come to as soon as he had re-examined his chain of reasoning?

7 In a certain community, when were the body parasites observed to leave a person?

8 What inference did the people draw from this? What was the correct explanation, and what error in reasoning does this illustrate?

9 Name (i) a favourable, (ii) an unfavourable example of the value of analogy in scientific research.

10 Name two other ways, mentioned in the passage, in which the reasoning process can be adversely affected. Give examples of each.

11 What advantages can an awareness of possible errors in thinking bring to the investigator?

12 Give other words or expressions which mean approximately the same as: non-specialist; more than; to be related to; to explain; as well.

Word Study

SYNONYMS

EXERCISE (a)

For each of the words appearing in column A below (all of which occur in the reading passage), choose a word from column B which means approximately the same:

A	B	A	B
to reason	complex	instance	later on
way	idea	to obtain	the opposite
frequently	to think	erroneous	case
complicated	to distinguish	the reverse	extremely
concept	in addition to	subsequently	roughly
to differentiate	to find out	to yield	to get
apart from	often	exceedingly	wrong
to discover	manner	approximately	to produce

(b)

Choose appropriate words from column B to replace the words in the following:

Apart from the experimental errors he is likely to make, an investigator sometimes finds that he has arrived at conclusions which are *approximately the reverse* of what he had expected. However, failure *to reason* in a logical *way* is *frequently* the reason why a series of *exceedingly* accurate experiments may appear to *yield erroneous* results, so this possibility should be checked in every *instance*. An additional difficulty is that words sometimes fail to *differentiate* clearly enough between the different aspects of a single *complicated concept*: the word 'average' is an *instance* of this.

WORDS WITH DIFFERENT FUNCTIONS

In addition to the fact that a word may have 2 or more totally different meanings (see Unit 8, Word Study section, a word may also have 2 or more different *functions*, i.e. it may operate as a noun, verb, and sometimes as an adjective as well. Thus *reason* is used both as a verb and a noun in l. 4. Similarly, *record* is used as a verb in l. 33 and as a noun in l. 35. The word *level* may be used as noun, verb or adjective, as in these sentences: The *level* of the liquid fell as the air pressure decreased; A building site has to be carefully *levelled* before work can begin; A *level* site was chosen for the new laboratory.[1]

Some common scientific words having more than one function are: alternate; approach; branch; chance; charge; collapse; conduct; cool; count; deposit; design; dip; drift; effect; fall; face; fit; forecast; influence; joint; jump; key; lag; mean; minute; multiple; net; obtain; peak; result; run; scatter; store; test; trace; track; value; waste.

EXERCISE (a) Look up the different functions and meanings of the above words in the Basic Dictionary, and incorporate as many as possible into sentences, using each word in the corresponding different functions.

(b) Look through the Basic Dictionary and note further examples of words having more than one function (i.e. those marked n. and v.; n. and adj.; n., v. and adj.)

WORD-BUILDING

1 The suffix *-able* (*-ible*). This suffix is added to verbs or nouns to form corresponding adjectives, e.g. *avoidable* (l. 3) meaning: which can be avoided; *unreasonable* (l. 42): which is not subject to reason; *valuable* (l. 63): which is of value; *visible* (l. 67): which can be seen.

NOTE: *applicable* (which can be applied); *destructible* (which can be destroyed); *intelligible* (which can be understood); *flexible* (which can be flexed or bent) and *soluble* (which can be dissolved).

EXERCISE Give the verbs corresponding to the following adjectives, and say what each one means:

breakable; calculable; convertible; reversible; comparable; divisible; predictable; transferable; identifiable; variable.

2 The prefix *non-*. This is freely added to adjectives and nouns to give the sense of *not being* or *not having*, e.g. *non-scientist* (l. 2) and *non-specialist* (l. 14), meaning: a person who is *not* a scientist or a specialist. Similarly the adjective *non-objective* means: which is *not* objective.

Some adjectives commonly used in scientific English with this prefix are: available; cumulative; effective; operable; pressurized; related; standardized; technical; transferable; uniform.

[1] It should be noted that occasionally the change in function may lead to a total change in meaning (unlike the examples already quoted): thus *mean* can be used as a n., v. or adj., but whereas the n. and adj. (in one of its two different meanings) are related in meaning, the word has a totally different significance when used as a verb.

3 The prefixes *re-* and *mis-*. These prefixes are freely added to verbs, and the adjectives and nouns formed from such verbs. *re-* usually gives the additional meaning of again, e.g. *re-examine* (l. 43), meaning: to examine again; *re-consideration* (l. 60): an additional or further examination. It can also have the secondary meaning of: back or against as in *react*, *return*, etc. *mis-* adds the meaning of: bad(ly) or wrong(ly), e.g. *misuse* (l. 32): wrong use; *to misapply*: to apply wrongly or badly, etc.

EXERCISE (a) Add *re-* to the following verbs, and use them in illustrative sentences:

arrange; assemble; distribute; integrate; organize; plot; value; straighten.

(b) From the words formed in **1**, **2** and **3** above, choose appropriate ones to fill in the blanks in the following sentences:

1 Many sick persons die every year from organic diseases which are at present non-....
2 When two or more elements react to form a compound, and this compound can then be separated back again into the original elements, the process is called a ...ible reaction.
3 In order to maintain the efficiency of a laboratory or workshop in the face of changing conditions, it usually has to be re-...ed periodically.
4 One of the characteristics which differentiates science from art is supposed to be that in the former knowledge accumulates progressively, whereas in the latter it is largely non-.... (Do you agree?)
5 The difficulty about using words like *accurate*, *hard*, *soft*, *hot*, *cold*, *long*, *short*, etc., is that their meanings are relative and therefore extremely ...able. In such cases the degree of accuracy, hardness, etc., must be indicated in quantitative terms.
6 If a member of a working group becomes sick or goes away and is not replaced, his work usually has to be re-...ed among the others.
7 Owing to lack of care, the specimens which the field-worker brought back to the laboratory were so badly preserved that most of them were not ...able.

Structure Study

PAST PERFECT TENSE

This tense is formed by *had* plus the Past Participle of the main verb,

e.g. l. 37: 'When he *had obtained* the data, he summed it up ...'

The main function of this tense is to indicate which of two or more *connected* actions in the past happened first. Thus in the example given above there are two related actions, i.e. *obtaining* the data and then *summing* it *up*. The *obtaining* happened before the *summing up*, so the first action goes into the Past Perfect Tense, while the second action is put into the Simple Past Tense.

Thus also ll. 43–44: 'The investigator himself realized (it was an unreasonable inference) as soon as he *had re-examined* his chain of reasoning.'

First came the re-examination (verb in Past Perfect), then came the realization of the mistake (verb in Simple Past).

Note that the connection between the events is often indicated by an adverb of time such as *when*, *as soon as*, *after*, *before*, etc.

EXERCISE

In the following sentences, put the verbs in brackets into the appropriate tense, i.e. Simple Past or Past Perfect:

1 As soon as the investigator (discover) his error, he (make) the necessary corrections to his data.

2 The authorities (begin) to attack Galileo as soon as he (publish) his new cosmological theories.

3 The accident occurred because the new aircraft (be put) into operation before it (be tested) sufficiently.

4 The medical researcher (refuse) to allow his new drug to be used until he (complete) all the necessary trials.

5 In spite of the fact that the young psychologist realized he (use) an inadequate sample in his investigations, he (allow) his results to be published.

6 During an experiment, a certain piece of metal bent under the strain to which it (be subjected). When the engineer (re-straighten) it, he (find) that it (lose) some of its original strength.

CONDITIONAL SENTENCES

These are sentences which express what happens if certain conditions are fulfilled. They therefore consist of two parts (which can occur in any order), i.e. a 'Condition' clause—usually introduced by *if* or *unless*—and a 'Result' clause. Each of these clauses may have different forms, which are reflected in the tenses used, which indicate the writer's attitude to:

(i) the chances that the conditions will be fulfilled,
(ii) the chances that the results will follow.

It is therefore obvious that a fairly large number of different types of conditional sentence are possible, but here we are only concerned with those that are commonly used in scientific English:

1 *If* (*unless*) plus the PRESENT tense for the verb expressing the Condition (indicating that the writer thinks that there is *no obstacle* to the condition being fulfilled, or that it can be or is sure to be fulfilled), connected to another PRESENT tense in the Result clause (indicating that the writer thinks that, given the conditions postulated, the result is normal or usual),

e.g. 'If we OBSERVE the actions of men, we FIND ...' (ll. 1–2),
If the supply of heat in a closed system INCREASES, the temperature RISES.
Unless care IS TAKEN, many avoidable mistakes ARE MADE.

2 Sometimes the writer wishes to indicate other attitudes to the Result: in these cases he uses an appropriate anomalous finite, usually WILL (emphasizing the futurity or inevitability of the result), MAY (indicating that the result is only possible) and SHOULD (indicating advisability),

e.g. If the pressure RISES above *p* lb/sq. in., structural failure MAY occur; if this limit is reached, therefore, the pressure SHOULD immediately be reduced and the experiment begun again.

3 *If* (*unless*) plus the PAST tense for the verb expressing the Condition (indicating that the writer thinks that the condition is not certain or is unlikely to be fulfilled—it is only a possibility or hypothesis), connected to the anomalous finites WOULD (indicating that the writer thinks that, given the conditions postulated, the result is inevitable) or MIGHT (indicating that he thinks the result is only possible, not certain),

e.g. 'If a researcher PERFORMED an experiment (hypothesis), ... it MIGHT be a failure (possibility only)' (ll. 18–20),
If all the ice in the world MELTED (possibility only), the level of the sea WOULD rise (inevitable result) by about 250 ft. (How many metres is this?)

EXERCISE

Complete the following conditional sentences, putting the verbs in italics into the right tense in accordance with the instructions given in brackets in each case:

1 If the body temperature *rise* (can happen) above 106°F, death frequently *occur* (usual result).

2 Unless the scientist *learn* (no obstacle) to think logically, he *commit* (inevitable) many mistakes.

3 If changes in laboratory conditions *occur* (can happen), the results of experiments *be affected* (possibility only).

4 If governments *spend* (possibility only) more money on education and scientific research, more rapid progress *result* (inevitable).

5 If intelligent beings *exist* (hypothesis) in other parts of the universe, they *try* (possibility only) to communicate with the earth.

6 Unless scientists and technologists constantly *question* and *re-examine* (hypothesis) established concepts and procedures, scientific progress *slow down* (inevitable) or stop.

TWO 'CAUSE-AND-RESULT' STRUCTURES

1 *Too hot to live on* (l. 56). This type of construction (TOO plus adjective or adverb plus verb with TO) expresses the result of an excess of something; in this case it is equivalent to: excessively hot, so that the parasites are not able to live on the body. Other examples are: This piece of metal is TOO hot TO hold (i.e. so hot that it cannot be held); That piece of apparatus is TOO heavy TO move (i.e. so heavy that it cannot be moved); That problem is TOO difficult *for me* TO solve (i.e. so difficult that *I* cannot solve it), etc.

2 *The straighter he thinks, the more successfully he will be able to perform his duties as a citizen* (l. 87). This structure (linking two comparatives) expresses the way in which a change in one thing causes a corresponding change in another related thing. Thus in the example quoted, increased power of logical thought leads to a corresponding increase in success as a citizen. Other examples are:

THE HIGHER the temperature, THE FASTER the speed of molecules; THE MORE CAREFULLY we work, THE BETTER the results; THE GREATER the degree of industrialization in a country, THE HIGHER the standard of living, etc.

Note that the construction is often used to state general principles.

SUBSTITUTION TABLES

I. Past Perfect Tense

1	2	3	4	5	6	7	8
The final theory A tentative hypothesis The new model	was	developed put forward modified	when after as soon as once	all the	data facts evidence information	had been	collected obtained analysed plotted

II. Conditional Sentences

A

1	2	3	4	5
If	the temperature rises, more pressure is applied, environmental conditions are altered, loads increase,	the alloy the metal the sample the material	necessarily inevitably always	undergoes a change of crystal structure loses its original properties develops new properties loses strength

B

1	2	3	4	5	6	7	8	9
Unless	reliable instruments	are	used	in	experiments,	misleading results	will certainly	be obtaine
	adequate controls		applied		field work,	inaccurate data	may possibly	
	effective procedures		put into operation		every type of research,	wrong conclusions		
	suitable techniques							
	critical judgement	is						
	logical thought							
	a careful plan							
	strict objectivity							

C

1	2	3	4	5	6	7
If	politicians	employed	efficient techniques,	skilled manpower	would certainly	increase
	governments	used	reliable procedures,	crops	might possibly	
	the state	applied	effective methods,	industrial output		

D Could (conditional form, equivalent to *would be able to*)

1	2	3	4	5	6
If	larger facilities	were available,	investigators	could	work more efficiently
	better equipment		scientists		improve existing conditions
	more capital		researchers		develop new sources of power

NOTE: In conditional sentences of the above type, the correct form of *to be* is *were* for both plural and singular nouns (column 3).

Additional exercise: Repeat the above tables putting the Result clause at the beginning of the sentence and the Condition (*If* or *Unless*) clause at the end, e.g.: The alloy necessarily undergoes a change of crystal structure if the temperature rises. (Table A.)

Discussion and Criticism

1 What are the difficulties connected with the word *average*? (You may have to find out the meanings of the words *mean*, *median* and *mode* and refer to a book on statistics to clarify this.)

2 Give examples from general science or from your own discipline of how misunderstandings and errors can be caused by words.

3 Explain the error in logic made by the young sociologist (ll. 39–41); a simple diagram may help you to do this more clearly.

4 Assuming you were the medical researcher concerned in the properly-controlled investigation referred to in ll. 53–56, describe how you carried it out.

5 Give additional examples of the 3 obstacles to reasoning outlined in ll. 57–60.

6 Do you know any further types of crooked thinking NOT mentioned in the passage? If so, explain them and give examples.

7 Discuss the following, and explain the errors in each case:

(a) Education implies teaching; teaching implies knowledge; knowledge is truth; the truth is the same everywhere. Hence education should be the same everywhere.

(b) X is one of the best politicians in the Government. Science is being applied to politics to an increasing degree. Therefore X is the best person to be in charge of the Government's scientific research programme.

(c) 'I am the Master of this College,
And what I don't know isn't knowledge.'

(d) Notice put up in an industrial research laboratory: 'Keep your bright ideas to yourself, and do not disturb your superiors: they have enough work of their own. Remember—99% of bright ideas turn out to be wrong!'

(e) Physiologists have established that the temperature at which man is able to work best is 62°F. The average temperature of the centre of the Atacama desert is 61·8°F.

(f) 'Industrialization for underdeveloped countries? A totally wrong solution! Look at the example of Y-land: that country started to industrialize 10 years ago and it has had nothing but political and economic difficulties ever since!'

(g) 'What is it that separates the living from the non-living? It is, quite simply, Nature's great principle of *vitalism*, recognized by the leaders of scientific thought of all nations under different names, from Bergson's *elan vital* to Shaw's *Life Force.*'

(h) Either a thing exists or it does not exist, so there cannot really be such things as *unreal numbers* or *anti-matter*.

BIBLIOGRAPHY

FEARNSIDE AND HOLTHER *Fallacy*, Spectrum Books.
THOULESS *Straight and Crooked Thinking*, Pan Books.
EMMET *The Use of Reason*, Longmans.

The techniques of problem-solving are discussed briefly in a section of the last-named book above, and at length in:

POLYA *How to Solve it*, Anchor Books.

There are a number of collections of logical and inferential puzzles available, e.g.

WYLIE *101 Puzzles in Thought and Logic*, Dover.
PHILLIPS *My Best Puzzles in Logic and Reasoning*, Dover.

Unit 10

Revision of material appearing in Units 6–9

SCIENCE AND THE FUTURE

In preceding (previous) units, we have examined briefly some of the characteristics, methods, effects and problems of present-day science. At this stage it may be worth considering a few of the ways in which it may develop in the near future, i.e. the next decade or so.

To begin with, we can expect applied science to produce a vast (huge, enormous) increase in entirely new synthetic products of all kinds. These will range from light-weight, high-strength materials for use in the many specialized branches of engineering, to drugs and chemicals with a greatly-increased selectivity which can be used in medicine and agriculture. However, in this latter case in particular, it may be predicted that the wide-spread application and combination of new and more complex products will give rise to unexpected inter-reactions or side-effects. For this reason, greatly intensified programmes of research will be required in order to discover and eliminate the harmful results of such combinations.

Another point is that the rapid expansion of industrialization throughout the world must inevitably lead to a progressive exhaustion of natural resources. If we wish to counter-balance these losses to some extent, we shall have to follow two main courses of action: (a) much greater efforts will have to be applied (devoted) to conservation, particularly of such items as soil, water, fuels and minerals; (b) more efficient methods of exploitation and utilization will have to be developed.

In the more developed countries, the automatization of industry (automation) will lead to a high degree of efficiency in the production of manufactured goods, and is likely to have far-reaching social effects. For instance, workers will need to be more highly trained and more flexible—they will probably have to be capable of changing (shifting) from one skilled job to another—and they will also have more free time, as they will work fewer hours per day. This in turn will necessitate a considerable expansion and re-orientation of education. Another result of automation should be to accelerate (speed up) the accumulation of surplus capital, which could then be made available for the purpose of assisting the emerging countries to solve some of the problems of underdevelopment. It should, however, be borne in mind that this process itself might involve a chain of difficulties, in this case of a political nature.

In general, the application—or misapplication—of science and technology in all fields is certain to affect the structure of society as a whole. This will remain true whether we are dealing with the application of psychology to advertising and political propaganda, or engineering to the mass media of communication, or of medical science to the problems of overpopulation or old age. This could lead to the development of a special discipline, whose job would be to estimate (evaluate) the social consequences of all major research and development (R and D) projects before they are put into large-scale operation. It should here be pointed out that one of the most powerful trends in present-day science is for separate branches to converge and form inter-related groups of studies. If this trend continues, it may in fact lead to the emergence of an entirely new type of scientist, i.e. the multi-disciplinary co-ordinator. 55

As we have previously seen, international co-operation has become greatly intensified in recent years, and this tendency will doubtless become even more strongly marked in the future. It is therefore likely that the scientific efforts of individual countries will tend to be unified and co-ordinated by supra-national entities, and the more this is done, the greater the probability that supra-national governments will eventually be set up. 60

National governments, also, will be brought into closer and closer contact with science. To quote only one reason, the State will have to provide an increasingly large proportion of the money spent on scientific investigation: it can therefore be expected to play an increasingly important role (part) in the planning of R and D programmes. It will also tend to determine one of the fundamental questions affecting science in the future, viz. the percentage of the funds which are made available for basic research, and the percentage allotted to development projects. From another point of view, the cumulative use of science in government must have the overall effect of greatly extending the control of the State over the ordinary citizen. All these factors, and many other related considerations, should stimulate a great deal of re-thinking on this subject, the results of which could bring about (cause) a scientific revolution in politics, or a political revolution in science—or both. 65 70 75

Comprehension

1 What can be expected to be one of the first consequences of applied science in the future? Give examples.

2 What unexpected effects may the widespread application of new drugs and chemicals have? Why?

3 Why are natural resources rapidly becoming exhausted? What items are particularly outstanding in this respect?

4 Name two methods by which this process can be counterbalanced.

5 What will be the initial (first) consequences of automation? Describe (i) a possible social effect, (ii) a possible political effect of this process.

6 Why might the development of a special social-science discipline be needed in the future? Describe the work of this discipline.

7 What do you understand by the term *multi-disciplinary co-ordinator*?

8 Why might scientists of this type be required in the future?

9 Name some of the possible consequences of international co-operation in the next decade or so.

10 Name some of the ways in the future in which (i) national governments will affect science. (ii) science will affect national governments.

11 What further results may this interaction have?

12 Give other words meaning approximately the same as: huge; to accelerate; R and D; to change (i.e. to change *position*, in particular); role; to evaluate; to cause.

Word Study Revision

WORD-BUILDING DEVICES

EXERCISE (a) The reading passage contains numerous examples of prefixes and suffixes which have been presented in previous units. Pick these out, and give the meaning in each case.

(b) Each suffix in List A can be connected to three words in List B. Form these words and give the vernacular for each.

A						
	-ify	-ly	-less	-ent	-ence	-or
	-ness	-able	-ible	-ant	-ance	-er

B.				
	use	investigate	differ	liquid
	effective	boil	reverse	possible
	identify	diverge	exist	transfer
	simple	count	assistant	direct
	reason	predict	resist	motion
	convert	solid	research	persist
	divide	absolute	interfere	maintain
	distant	hard	react	signify
	evident	meaning	produce	perform

(c) Using words formed in Exercise (b) above, fill in the blanks in the following sentences:

1 An eroded area of land is u...less for agriculture.

2 The d...ance between the earth and the moon is about 385,000 km. (How many miles is this?)

3 Most volcanic lavas s...ify at 800–1200°C.

4 Diamonds, which are a form of carbon, have many industrial applications, due to their h...ness.

5 To some people, the concept of anti-matter is me...less.

6 Mass is co...ible into energy.

7 The atomic r...or was developed with the as...ance of re...ers from many different countries.

(d) Fill in the blanks in the following sentences, using the prefixes *under-*, *over-*, *mis-*, *re-* and *non-*:

1 When the engineer ...straightened the bent piece of metal, he found it had lost part of its strength.

2 Researchers tend to ...estimate the importance of subjective factors in an investigation.

3 These factors may cause a scientist to ...stress irrelevant details.

4 It is difficult to explain scientific processes in ...technical language.

5 One of the most serious social problems of our age is that of the ...use of science by governments.

6 The result of ...production is a surplus.

7 The bulk of the population in the emerging countries is ...nourished.

COMPOUND NOUNS

EXERCISE

Replace the italicized phrases by compound nouns:

1 The Industrial Revolution led to a radical change in *methods of production.*

2 *The mechanics of fluids* is a specialized branch of physics.

3 *Prices for the produce of farms* rise when there is a *failure of the crops.*

4 Nuclear energy has been utilized by some of the latest *plants for the supply of electric power.*

5 Gas is one of the by-products of *the industry concerned with the refining of oil.*

6 Earthquakes give rise to several different kinds of *waves of shock.*

7 Several countries are carrying out *plans in the field of economics which are to be carried out in units of five or ten years.*

PHRASAL (PREPOSITIONAL) VERBS

EXERCISE

Complete the verbal forms in the following:

1 At normal pressures, water t...s into ice at about 0°C. (How many °F is this?)

2 Political instability in one area may l... to serious disturbances in other parts of the world.

3 In order to s... up industries, trained manpower, as well as sufficient capital, is needed.

4 The researcher should always b... in m..., when c...ing out an investigation, the fact that experiments frequently t... out to have totally different results from the ones expected.

5 Although there are dozens of special alloys used in engineering, most of them are m... up of different proportions of only 3 or 4 metals.

Structure Study Revision

TENSES

EXERCISE (a)

Put the verbs in brackets into the correct tense (Simple Present, Present Continuous, or Present Perfect):

1 Unit 9, which we just (complete), (demonstrate) the importance of straight thinking.

2 In the last few decades man (succeed) in utilizing the atomic nucleus as a source of energy.

3 At present, a number of research teams (look) for a new approach to the synthesis of proteins.

4 Many developing countries (retain) old and inefficient economic structures which (prevent) them from making the best use of modern agricultural and industrial methods.

5 The new machine, which (break down) several times recently (run) smoothly now.

6 Some types of steel (melt) at c. 1200°C.

7 In the last few years, chemists (develop) hundreds of synthetic materials.

(b)

Put the verbs in brackets into the correct tense (Simple Past or Past Perfect):

1 Success (come) after we (perform) the experiment 20 times.

2 The engine (break down) before it (complete) all the reliability tests.

3 After the properties or characteristics of the new material (be worked out) theoretically, they (be tested) experimentally.

4 The field studies we (undertake) last year (be) very successful because they (be) carefully planned beforehand.

5 The construction of the bridge (begin) as soon as the engineers (make) the necessary calculations.

6 It (be) necessary to repeat the experiments because some of the original data (be lost).

7 Unfortunately, it (be) difficult to repeat the experiments since the original conditions (change) in the meantime.

CONDITIONAL SENTENCES

EXERCISE

Complete the following conditional sentences by putting the italicized verbs into the correct tense in accordance with the instructions given in brackets:

1 If observations or other data are inaccurate or insufficient, or if the reasoning used is false, our experiments *be* (inevitable) failures.

2 If the present trend towards increased specialization continues, inter-disciplinary communication *become* (possibility only) more difficult.

3 In underdeveloped countries, output could be increased if working conditions *be improved* (hypothesis only).

4 Many aspects of the universe *seem* (usual result) meaningless if we try to give a common-sense explanation of them.

5 Unless laboratory conditions are kept constant, many pieces of apparatus *become* (possibility) inaccurate.

6 If additional funds were made available for development projects at any time, more money *need* (inevitable) *to be allotted* to basic research as well.

7 If possible harmful effects of new products and processes are to be avoided, research programmes to evaluate their results *be set up* (advisability).

INFINITIVES AND *-ing* FORMS

EXERCISE

Put the verbs in brackets into either the Infinitive or '*-ing*' form, as required:

1 We have not yet got used to (use) the sea as a major source of food.

2 The process of (solve) a problem by analogy may often give good results.

3 It is sometimes difficult (distinguish) between cause and effect.

4 The geologist helped the engineer (calculate) the stability of the building by (give) an accurate description of the soil.

5 One of the machines stopped (work) and caused a blockage in the assembly line.

6 A good scientist enjoys (try) (solve) difficult problems.

7 Planners tend (underestimate) the importance of psychological factors in underdevelopment.

ANOMALOUS FINITES

EXERCISE

In the following sentence, fill in the blank with the appropriate anomalous finite which conveys the idea of:

1 compulsion,

2 moral obligation,

3 physical or mental ability,

4 possibility,

5 more remote possibility,

6 moral obligation—(different from 2 above,)

7 necessity in the past.

They ... test the apparatus.

Discussion and Criticism

1 Explain clearly, and as fully as possible, any examples known to you of unexpected side-effects (both beneficial and harmful) which new products and processes have had.

2 Give clear illustrations of the exhaustion of natural resources.

3 How can natural resources be conserved? Give specific examples.

4 What is your opinion with regard to the possible consequences of automation outlined in ll. 26–40? Give good reasons for your point of view.

5 Discuss the examples given in ll. 43–47 of the ways in which the application of science is affecting the structure of society. Give further illustrations of your own.

6 What are some of the ways in which you think your own discipline will develop in the future?

7 If you were the Minister of Science for your country, in what proportions would you divide the funds, as between basic science and development projects? Give good reasons for your answer.

8 What do you understand by the last sentence of the passage? Explain as fully as possible.

BIBLIOGRAPHY

CALDER (ed.) *The World in 1984*, 2 vols, Penguin.
GABOR *Inventing the Future*, Penguin.
SOULE *The Shape of Tomorrow*, Signet.
BAGRIT *The Age of Automation*, Penguin, Mentor.
HARRISON BROWN *The Challenge of Man's Future*, Compass.
HUXLEY *Brave New World Revisited*, Bantam.

Unit 11

(General Revision Unit)

THE ROLE OF CHANCE IN SCIENTIFIC DISCOVERY

Nearly a century and a half ago, a Danish physicist, Oersted, was demonstrating current electricity to a class, using a copper wire which was joined to a Voltaic cell. Amongst the miscellaneous apparatus on his demonstration bench there happened to be a magnetic needle, and Oersted noticed that when the hand holding the wire moved near the needle, the latter was occasionally deflected. He immediately investigated the phenomenon systematically and found that the strongest deflection (deviation) occurred when he held the wire horizontally and parallel to the needle. With a quick jump of imagination[1] he then disconnected the ends of the wire and reconnected them to the opposite poles of the cell—thus reversing the current—and found that the needle was deflected in the opposite direction. This chance discovery of the relationship between electricity and magnetism not only led quickly to the invention of the electric dynamo and hence to the large-scale utilization of electric energy, but forms the basis for modern electro-magnetic field theory, which is now an extremely valuable tool in both macro- and micro-physics.

The above story illustrates the part played in scientific discovery by chance (accident). Again, about 20 years ago a group of British bacteriologists and biochemists working in agricultural research were carrying out investigations into substances of organic origin which could be used to stimulate plant growth. One of the approaches they used consisted in studying the nodules (small round lumps) found on the root-hairs of certain plants, and which contain colonies of nitrogen-forming bacteria. Working on the hypothesis that these bacteria manufactured a substance which stimulated the nodule-forming tissue, the investigators eventually succeeded in isolating this substance. However, when they then tested it on various other plants, they found—quite contrary (opposite) to their expectations—that it actually prevented (inhibited) growth. Further systematic investigation showed that this toxic (poisonous) effect was selective, being much greater against dicotyledon[2] plants, which happen to include the majority of weeds, than against the monocotyledons[2], which include the grain crops and grasses. The researchers thus realized that they had discovered a powerful selective weed-killer: they continued their research, using inorganic compounds of related chemical composition, and in this way laid the foundations of a technology which is of the greatest value in present-day agriculture.

Another well-known instance of the role of chance is connected with the discovery of penicillin by Fleming. This medical researcher had been investigating some pathogenic (disease-causing) bacteria, and after being absent from his laboratory for

[1] Or *intuition*, as it is often called when it produces successful results.
[2] Usually *dicots* and *monocots* respectively in U.S.A.

some days found on his return that one of the culture dishes in which colonies of the bacteria were growing had been contaminated by a colony of another organism, a mould of penicillium *spp*. He was going to throw the dish away when he noticed that the penicillium colony was surrounded by an area completely clear of the pathogenic bacteria. He immediately realized that the penicillium must have manufactured a substance which had broken down (disintegrated) the pathogenes. He then isolated this substance, which turned out to be the most powerful agent yet discovered against bacteria causing a number of dangerous and widely-spread diseases.

Apart from demonstrating the way in which chance may lead to scientific discoveries of primary importance, an analysis of the three cases outlined above may be useful in showing *how* a successful worker utilizes these accidental opportunities. The first point to notice is that although in all cases the key phenomenon produced results which were both unexpected and—in the last two cases—even apparently disadvantageous, the scientists invariably reacted in an extremely positive manner. The refusal to be disturbed or disorganized by unexpected or apparently adverse occurrences, but, on the contrary, to be stimulated by them, has in fact been a marked (strong) characteristic of successful investigators.

Secondly, we note that in the first and third cases the phenomena were very slight and might easily have escaped notice, whilst in the second case they produced a negative result. From this we might deduce that a superior capacity for observation is also a property of outstanding researchers. On this point, however, a psychologist would probably tend to disagree. He would point out that observation or perception is a concept which refers not so much to acuteness of sight, hearing, etc., or to the care with which they are applied, as to the ability to relate phenomena to a complex network of previous experiences and theories, i.e. to a meaningful frame of reference. In other words, an observer who lacks such a frame of reference will be unable to realize the significance of certain phenomena even though his senses may 'experience' them, and so he may fail to observe them. This can be illustrated by the following example:

At the end of last century, an American chemist, Hillebrand, was using a recently-developed instrument, the spectroscope, to analyse the gas given off by a certain mineral when treated with acid. This instrument works on the principle that each individual substance emits a characteristic spectrum of light when its molecules are caused to vibrate by the application of heat, electricity, etc; and after studying the spectrum which he had obtained on this occasion, Hillebrand reported the gas to be nitrogen. At this same time, another scientist, Rayleigh, happened to be investigating the anomalous fact that nitrogen obtained from the air appeared to be heavier than that obtained from other sources, e.g. ammonia (NH_3). Rayleigh repeated Hillebrand's experiment and, immediately noticing that the spectrum showed several bright lines which were additional to

those typical of nitrogen, went on to discover the rare gases argon (A) and helium (He). Why had Rayleigh observed these
100 extra lines whereas (while) Hillebrand apparently had not? Part of the answer seems to be that the former already possessed a frame of reference which included the possibility that a different sort of N might exist; he was therefore extremely sensitive to any apparent anomaly in the behaviour of this element.
105 Hillebrand lacked this concept, and was therefore unable to *see* the slight deviant reaction of the gas he assumed he was dealing with.

This dual (double) quality of being sensitive to, and curious about, small accidental occurrences, and of possessing a frame
110 of reference capable of suggesting their true significance, is probably what Pasteur meant when he said 'Chance benefits only the prepared mind.' Nevertheless, it is clear (plain, obvious) that these qualities alone, even when joined to those mentioned previously, are not necessarily sufficient to ensure success: an
115 indispensable factor in all the discoveries quoted above was careful and systematic experimentation. We may therefore conclude that it is the capacity to plan and undertake such experimentation which finally allows the investigator to make the most of his luck —if it comes.

Comprehension

1 What was the accidental phenomenon which Oersted noticed and investigated?

2 How did he make the needle deviate to the *opposite* direction to that of its original deflection?

3 What forms the basis of modern field theory?

4 What substance did the British agricultural researchers succeed in isolating?

5 Why were its effects on dicots of great interest to the investigators?

6 What are pathogenes?

7 What evidence did Fleming find which led him to assume that the penicillium broke down the pathogenes?

8 Besides illustrating the role of accident in scientific investigation, what else can we learn from the cases quoted?

9 Describe a strong characteristic of successful researchers which is demonstrated in each of the examples given.

10 What deduction regarding researchers might a psychologist disagree with?

11 Describe the concept of *observing*.

12 What is a spectroscope?

13 Why was Rayleigh interested in nitrogen?

14 What phenomena led him to discover the inert gases argon and helium?

15 What did Pasteur probably mean by *the prepared mind*?

16 What additional capacity is usually necessary for the successful exploitation of accidental occurrences?

17 Give words meaning approximately the same as: to disintegrate; opposite; chance; marked; imagination; to inhibit; plain; deviation; whereas.

Word Study

WORD-BUILDING

The prefix *dis*—. This is attached to words, mainly verbs and their derived adjectives and nouns, to give a negative or opposite meaning, e.g.: *dis*connect (l. 10) the opposite of *to connect*; *dis*organized (l. 65), meaning *not organized*, etc.

EXERCISE (a)

Add the prefix *dis*- to form opposites of the following:
(nouns): ability; advantage; appearance; order; use.
(verbs): agree; like; prove; integrate.
(adjective): similar.

(b)

Fill in the blanks in the following with appropriate words formed in Exercise (a) above:

1 When one thing is different from another, the two things are said to be ....

2 In many cases, economic instability may lead to political and social ....

3 Hillebrand's assumption regarding the nature of the gas he had studied was ...ed by Rayleigh's investigations.

4 It is well known that atrophy, i.e. the wasting away of certain mental and physical characteristics, is caused mainly by ....

5 When an integrated organism or system breaks down or is split up into its separate parts, it is said to ....

STRUCTURAL WORDS—MODIFYING CONNECTIVES

However (l. 30), *thus* (l. 37), *when* (l. 49), *apart from* (l. 57), etc., are examples of a very important class of words with a dual function, i.e. to connect the different parts of a statement and at the same time to modify its total meaning in some way. These should now be revised by reference to Part II of the Basic Dictionary (2. Modifying Connectives).

EXERCISE

Read the following passage carefully, making sure you know the meaning of the modifying connectives (in italics), and complete the alternatives in brackets:

Although (t-----, i- s---- o- the fact that) many complex problems will face the human race in the future, it is reasonable to assume that the application of the techniques of present-day science will in time (eventually) provide the solution to most of them.

To take (consider) only one of these problems—that of undernourishment—it can reasonably be predicted that one of its main causes, i.e. over-population, will eventually be solved by the widespread use of devices for controlling human fertility, *provided that* (s------ t- the proviso that) social customs and laws are relaxed sufficiently to allow this. At present this solution is *clearly* (o-------) limited *because of* (o---- t-, o- a------ o-) these social considerations.

Another way of approaching the problem is to increase the food supplies available: this, *also* (t--), can be done through (by means of) the scientific methods we have at our disposal now, e.g. by raising the yield of crops and animals. This is possible by the introduction of genetically improved varieties and by the use of fertilizers and weed-killers, *besides* (a- w--- a-) by the extension of more scientific methods of cultivation. *Moreover* (f----------), *inasmuch as* (i- s- f-- a-) these measures do not usually involve interference with existing religious or social patterns, they are more likely to be put into immediate operation than the method mentioned previously, viz. birth-control.

Notwithstanding (i- s---- o-, reg------- o-, ir---------- o-) what has been stated above, however, it may be doubted whether the measures referred to will be sufficient to prevent the situation from deteriorating (becoming worse)—in the near future, at least—unless some totally new factor appears. *Nevertheless* (h------, b--), some writers consider that it is fairly probable that such factors will *indeed* (i- f---) emerge, in view of the exponential growth of science, including the social sciences. *Otherwise* (i- n--), the outlook for the human race will indeed be dark.

REVISION—COMPOUND NOUNS AND NOUN PHRASES

In Unit 6 we studied the compound noun structure and its uses. There are a number of examples of these in the Reading Passage of the present unit, e.g. *copper wire* (l. 2), *demonstration bench* (l. 4), etc. Pick these out and explain what each one means.

Structure Study Revision

TENSES

EXERCISE (a)

Put the verbs in brackets into the correct tense, either Simple Present or Simple Past (note that some irregular verbs are included):

1 People sometimes (confuse) cause and effect.
2 In the case recorded in Unit 9, the young sociologist (draw) the wrong conclusion from his data.
3 People generally (keep) oil in large storage tanks.
4 Fleming (see) that the penicillium colony was surrounded by a clear space.
5 The hotter a substance (become), the faster the movement of the molecules.
6 Torricelli (put forward—i.e. develop) a theory of air pressure over 300 years ago.

7 People (apply) scientific knowledge to practically all fields of human activity, nowadays.

8 This recently-developed instrument is Japanese: they (make) it in Japan.

9 The faster an aircraft (fly), the greater the stress and strain on the materials.

10 We generally (use) the exponential notation when we (write) down numbers in scientific work.

(b) Some of the sentences in the exercise above would more usually be put in the Passive. Find these, and put them into the corresponding Passive form.

-ing FORMS

EXERCISE

Replace the expressions in italics with an appropriate *-ing* form in each case:

1 The force *that links* atoms together to form molecules is called the chemical bond.

2 Technologists are concerned in *the application of* pure science to practical affairs.

3 Air-speeds *that exceed* several times the speed of sound have been reached by modern types of aircraft in level flight.

4 A good investigator is interested in *the explanation of* the phenomena around him.

5 Engineers in many parts of the world are engaged in *the development of* machines capable of exerting extreme pressures on materials.

6 Celestial bodies, i.e. bodies in space *that revolve* round (around) larger bodies are called satellites.

7 New electronic devices are available nowadays to help the scientist in *the computation of* complex calculations.

8 The blood *that circulates* in our veins and arteries is a mixture of several substances.

9 A type of radiation *which consists* of light and sub-atomic particles is released by radioactive elements.

10 The efficiency of a power supply system can be calculated by *a measurement of* the difference between the potential intake (input) of energy and the actual output.

-ing FORMS AND THE INFINITIVE

EXERCISE

Put the verbs in brackets into either the Infinitive or the *-ing* form, as necessary:

1 A good scientist enjoys (solve) problems.

2 Polar species of plants or animals seldom get used to (live) in warmer climatic zones.

3 Industrialization enables countries (raise) their standards of living, but it is also necessary (improve) methods of agriculture as well.

4 An outstanding politician is not likely (succeed) in science. (Do you agree? Is the converse necessarily true?)

5 We ought (write) clearly when (record) our observations.

6 It is very difficult for a single scientist (cover) all the aspects of even one discipline.

7 In order to establish a series of wide and fertile frames of reference, a scientist should (develop) an interest in other disciplines beside his own.

8 Before (publish) his results, the wise investigator always asks other scientists (work) in the same field (check) their accuracy.

9 (Begin with), a young scientist often tends (overlook) the importance of (plan) his work carefully before (begin) it.

10 (Take) correct measurements, it is necessary (use) accurately-adjusted instruments.

IMPERSONAL EXPRESSIONS

EXERCISE

Repeat the following sentences aloud, putting the verb *be* in the correct tense (Present, Past, Future, Present Perfect):

1 Rayleigh (be) interested in Hillebrand's experiment because it (be) plain to him that the latter had used a new source to obtain nitrogen from. He thereupon decided that it (be) necessary for him to investigate the N obtained from this source.

2 During the next decade or so, it (be) difficult to provide enough food for the rapidly-increasing world population.

3 After we had completed a series of crucial experiments last week it (be) clear that our original hypothesis was incorrect (wrong).

4 Up to the present it (be) impossible to provide accurate long-term weather forecasts, though it (be) likely that a combination of meteorological satellites and computers will enable us to do so in the future.

5 Many people think that it (be) obvious that a greater proportion of the resources available should be allocated to basic research.

Discussion and Criticism

1 Describe in an orderly and accurate way an instrument or piece of apparatus which is used in the science you are studying. Then prepare clear and detailed instructions for its use, employing diagrams where necessary.

2 Do you agree with the remarks about *observation* given in ll. 74–83? If so, how do you think a scientist can acquire the 'wide frames of reference' required?

3 Analysing the cases outlined in the reading passage, do you think they illustrate other qualities necessary for successful investigation, apart from those mentioned in the text?

4 Describe any examples known to you which illustrate Pasteur's saying that 'Chance favours only the prepared mind'.

5 Re-read ll. 84–99. From the data given, can you form any hypothesis to explain why Rayleigh had found atmospheric nitrogen to be apparently heavier than N from other sources? How would you proceed to test this hypothesis?

6 In l. 17 there is a reference to *electro-magnetic field theory*, a specialized concept belonging to physics; l. 44 refers to *pathogenic bacteria*, a concept peculiar to medicine or bacteriology. Choose a similar specialized concept from the discipline you yourself are studying, and explain it so that a non-scientist (layman) could understand it.

BIBLIOGRAPHY

The books listed in the Bibliography to Unit 8 contain much interesting material on this subject, also:

WILSON *Introduction to Scientific Research*, McGraw-Hill.
LOTSPEICH *How Scientists Find Out*, Little, Brown.

Unit 12

(General Revision Unit)

THE SCIENTIST AND GOVERNMENT

We have already seen that science, besides affecting our whole environment—and hence the community as a whole—is becoming connected to an increasing extent with government itself (see Unit 10, last paragraph). As this is now beginning to modify the status (position in society) of the scientist himself, it may be worth while considering the subject in more detail at this stage.

To begin with, we may notice a number of additional factors which are accelerating this process. These are:

(a) *Defence Requirements:* The governments of most of the developed countries have always applied a large proportion of their total resources to the development of destructive apparatus and nowadays many of the newly-independent countries are doing the same. This apparatus and its use has now become exceedingly complex and requires the participation of large numbers of scientists and technologists. Moreover, many types of industry, including the largest, e.g. the aerospace industry, are strongly linked to defence requirements and again depend increasingly on scientific and technical personnel.

(b) *Economic Requirements:* Governments throughout the world act on the assumption that the welfare of their people depends largely on the economic strength and wealth of the community. Under modern conditions, this requires varying measures (degrees) of centralized control and hence the help of specialized scientists such as economists and operational research (O.R.) experts. Furthermore, it is obvious that the strength of a country's economy is directly bound up with the efficiency of its agriculture and industry, and that this in turn rests upon the efforts of scientists and technologists of all kinds. It also means that governments are increasingly compelled to interfere in these sectors in order to step up (increase) production and ensure that it is utilized to the best advantage: for example, they may encourage research in various ways, including the setting up of their own research centres; they may alter the structure of education, or interfere in order to reduce the wastage of natural resources or tap resources hitherto unexploited; or they may co-operate directly in the growing number of international projects related to science, economics and industry, such as the International Atomic Energy Agency, the European Iron and Steel Community or the various Common Markets. In any case, all such interventions are heavily dependent on scientific advice and also scientific and technological manpower of all kinds.

(c) *Social Requirements*: Owing to the remarkable development in mass-communications, people everywhere are feeling new wants and are being exposed to new customs and ideas (Unit 7, ll. 33–35), whilst governments are often forced to introduce still further innovations (things which are new) for the reasons given in paragraph (b) above. At the same time, the normal rate of social change throughout the world is taking place at a vastly accelerated speed compared with the past: for example, in the early industrialized countries of Europe the process of industrialization—with all the far-reaching changes in social patterns that followed—was spread over nearly a century, whereas nowadays a developing nation may undergo the same process in a decade or so. All this has the effect of building up unusual pressures and tensions within the community and consequently presents serious problems for the governments concerned. Additional social stresses may also occur because of the population explosion or problems arising from mass migration movements—themselves made relatively easy nowadays by modern means of transport. As a result of all these factors, governments are becoming increasingly dependent on biologists and social scientists for planning the appropriate programmes and putting them into effect.

(d) *Political Requirements:* Since defence, economics and social welfare constitute a very large proportion of the subject-matter of politics, it can easily be seen that science is already playing a major role in the political process. Not only this, but even in the field of action commonly considered to be purely political—e.g. the process whereby (through which) the politician obtains and retains the support of the people, or that of making rapid decisions after weighing the evidence from many different sources—it is again evident that he relies heavily on the scientist; in the first case, on the psephologist, social psychologist and mass-communications expert, and in the second, on the mathematician and computer scientist.

It follows from the above considerations that scientists, whether they like it or not, are becoming involved in government, and hence in politics itself, to a far greater extent than ever before. This fact raises several important questions:

First, how do these new circumstances affect the position in the community of the politician? For whereas in past ages, and even until recent times, the politician could act as a leader because he could reasonably claim to be an 'expert' on economics, war, human relations and the various other sciences that fall within the scope of government, he is now not only a non-expert, but has in almost every case been educated and trained in a non-scientific or even anti-scientific tradition. Yet when dealing with both long-term planning and short-term, day-to-day problems and arrangements he is, as we have just seen, forced to rely to a great extent on the advice and services of scientists. Thus it seems that a rapid change is occurring in the relative effectiveness—and hence of the relative positions in society—of politicians and scientists, and that the latter are moving more into the centre of the political process, possibly at the expense of the former. Second, how far is the scientist equipped to handle political and governmental affairs, apart from his specialized knowledge? Various general characteristics developed by his professional activities are certainly favourable, and some of these have already been mentioned in

100 Unit 1. Others are his use of the practical and experimental approach to problems rather than the dogmatic one, his preference for the long-term solution rather than the temporary and partial one, and—not least—his training in interdisciplinary and international co-operation. To offset these advantages, scientists
105 have so far (up to now) been on the whole unwilling to accept their new status and responsibilities in government. This rejection may be due to a mixture of reasons, e.g. because they are unaware of the situation or on the grounds that political involvement would interfere too much with their professional
110 work. Another objection arises from the marked difference commonly found between the type of education followed in the scientific disciplines and that of the humanistic branches of study. For if the education of the politician usually unfits him to understand science, the scientist too often lacks training in the
115 humane and social values of art, philosophy and human relationships. Although there are signs that this gap between the 'two cultures' is being bridged, much remains to be done.

Lastly, what are the special responsibilities, if any, of the scientists towards the community? It seems to us that on the one
120 hand he must make intensive efforts to give the ordinary citizen —and the politician—the means of evaluating the role of science in the modern world, since in the long run it is only the existence of a large body of well-informed and energetic citizens which can control abuses of governmental power; on the other hand, he
125 must take more trouble to prepare himself for his own growing role as decision-maker and administrator. Progress is being made on both these points, though mainly by the developed countries so far. As educators, increasing numbers of scientists are making an understanding of science more accessible to the
130 layman by means of simply-written books and radio talks, whilst in many countries the scientists are also the main force behind efforts to improve and expand the teaching of science and scientific method in schools and adult education courses. As far as the second requisite is concerned, progress has been slower,
135 for the reasons given above; nevertheless, one significant indicator for the future may be the widening scope and influence of the Pugwash Conferences, which are informal meetings held once or twice a year by leading scientists from all countries, in which science and its relationship to world affairs is discussed.
140 Another significant fact is that the necessity for establishing coordinated national science policies—now being put into effect in various countries—is bringing comparatively large numbers of scientists directly into government administration; this in turn is bringing about (causing) the evolution of a new type of scien-
145 tist, the scientist-administrator.

To conclude, it is clear that the whole world is passing through a social revolution in which a central part must be taken by scientists and technologists. But whether their efforts can be more effective than those of the traditional politician may depend
150 not so much on the present-day working scientists, but on the scientists now being trained, i.e.—you.

Comprehension

1 What circumstance is having the effect of altering the position of the scientist in society?
2 Name the four main fields of action which are accelerating the process of connecting the scientist with society.
3 Give two reasons why the scientist is taking a larger part in defence than hitherto.
4 What two sectors of the economy provide the main support for a country's economic strength?
5 Name some of the ways in which governments can stimulate the economic growth of their countries.
6 What sort of scientists (apart from economists) help in improving the economic situation of a country?
7 Name some of the main reasons why countries nowadays have to face unusual social stresses and strains.
8 What contribution does the scientist make to the purely political field of action of the politician?
9 Why does the position of the politician in society seem to be changing?
10 What advantages does the scientist bring to the job of administration? What are his disadvantages at present?
11 What are the reasons why scientists should popularize science? How is this being done?
12 What are the Pugwash Conferences?
13 Name some reasons for the evolution of the scientist-administrator.
14 Give other words and phrases meaning approximately the same as: so far; to increase; through which; position in society; to cause; a new idea or procedure.

Word Study

NEW VOCABULARY

EXERCISE

Complete the following with words taken from the reading passage. Approximate synonyms for the incomplete words are given in brackets.

One of the most important f--lds (areas) of government and intergovernment action, but probably the one to which the least amount of w---th (riches, quantity of resources) in money and research is being allotted at present, is the conservation of natural r------es (supplies). This comparative lack of interest is creating a m---r (very important) problem all over the world, since the w----ge (loss through misuse) of minerals, organic fuels and soil is increasing at a rapidly-accelerating r--e (speed). There are many reasons for the sp--d (rapidity) with which these resources are being exhausted. The first is the steep r--e (increase) in the world population, which shows no s--ns (indications) of levelling out. L--ked (connected) with this is the fact that the wants and re-----ments (needs) of people throughout the world are ex---ding (increasing) rapidly—in other words, there are more people wanting more of everything. Thirdly, many of the administrators or governments concerned l--k (do not have) the knowledge of how to use more ef----ive (efficient) methods of exploitation, or to utilize substitutes for the natural products which are becoming scarce. Lastly, even those who have the necessary knowledge may be slow to put into e----t (to apply) less wasteful methods, since they calculate that the harmful effects of their actions will fall, not on themselves, but on other people living in the future.

In view of the extremely long time needed to replace natural resources, it f----ws (results) that efforts to s--p up (increase the speed of) the application of more rational ways of using them is urgent. There is also great need to widen the sc--e (area of activity) of attempts to find substitute materials and new pr----ures (methods) for using old materials over and over again (repeatedly). Unless this is done soon, man will no longer control his en-----ment (surroundings) but be controlled by it.

WORD-BUILDING

The suffix *-ive*. This suffix is added mainly to verbs to form adjectives with the general meaning of: which *...s*, e.g. destruct*ive* (l. 11): which destroys; relat*ive* (l. 92): which relates, or expresses a relationship, etc.

EXERCISE

Look at the verbs in list A below and for each one pick out the corresponding adjective from list B. Use at least 5 of these adjectives in sentences of your own:

A expand; act; construct; compare; produce; exceed; effect; accumulate; imagine; explode; select; extend; describe; cause.

B productive; cumulative; explosive; active; effective; selective; expansive; extensive; comparative; descriptive; constructive; imaginative; excessive; causative.

NOTE: This suffix also forms adjectives from a limited number of nouns, e.g. *massive* (having a great amount of mass), *quantitative* (concerned with quantity), *intensive* (having a great deal of intensity), etc.

REVISION

1 *Words with the same appearance and function, but different meanings* (see Unit 8):

EXERCISE

From the following list, choose an appropriate word to complete each of the pairs of sentences given below, in which two different meanings of the word are illustrated:

a trace; frequency; a plant; to obtain; heavy; a solution; a sign.

1 (a) A relatively easy way of increasing agricultural production is to make use of better varieties of ...s.

(b) Countries which do not have sufficient supplies of mineral fuels or water resources for the production of electricity are now beginning to use atomic power ...s for this purpose.

2 (a) In some countries, the mathematical ... expressing division is ' : ', which, however, expresses ratio in English-speaking countries. The latter use ' ÷ ' for division.
(b) Contrary to popular belief, no ... of intelligent life has yet been detected on Mars.

3 (a) The completed graph taken from any type of self-recording instrument is often referred to as a ....
(b) Even when chemicals have been subjected to a very high degree of purification, they frequently retain ...s of other substances.

4 (a) One of the problems in modern high-speed engineering is that materials which have the necessary strength are often too ... to use.
(b) Recent geological evidence indicates that during some periods of the earth's history its magnetic field ceased to function. The earth would then have received a ... concentration of cosmic rays, which would probably have had the effect of modifying the evolution of living organisms.

5 (a) Radio signals can be received more clearly if they are transmitted at a very high ....
(b) In estimating the significance of an event, it is necessary to measure not only its intensity, but also the ... with which it takes place (occurs).

6 (a) When tackling a problem, an investigator should always try to reach a ... by the shortest and most efficient means. Modern statistical methods of planning experiments allow him to test several variables at the same time, thus cutting down (reducing) the time and costs spent on a series of experiments.
(b) A knowledge of the various chemical ...s used for staining and fixing organic material is an essential part of the training of a biological or medical researcher.

7 (a) It is usually very difficult to ... reliable statistics for use in the human sciences (sociology, economics, psychology, etc.) Hence the accurate compilation of relevant statistics is often a long and expensive preliminary step which has to be taken before a work of investigation can be effectively planned.
(b) High altitude and Polar research in the field are greatly hindered by the unfavourable climatic conditions usually ...ing in these regions.

2 *Words with the same appearance, but different functions and meanings* (see Unit 9):

EXERCISE (a) Complete the following sentences with words chosen from the list below. Each word should be used twice, in different ways: waste; test; minute; effect; lag; multiple; deposit; track.

1 Sedimentary rocks are made up of material which has been worn away from earlier rocks and ...ed on the bottom of seas or other low-lying areas.

2 After many careful ...s in the laboratory and workshop, the prototype finally broke down owing to metal fatigue.

3 In scientific investigation, much time and effort may be ...ed owing to lack of agreement about the meanings of words.

4 The movements of solutions in living organisms can be traced by putting ... quantities of radioactive elements into the subject and tracing them by means of radiation counters.

5 In many underdeveloped regions, food production ...s behind population growth.

6 The behaviour and interactions of sub-atomic particles can be studied by means of the ...s they leave in a highly saturated gas (e.g. as in a cloud chamber).

7 Certain areas of the Middle East are very rich in oil and many ...s of this valuable fuel are being intensively developed.

8 Angles can be measured in terms of degrees, ... s and seconds.

9 In many cases of mental disorder, a cure may be ...ed by means of methods which involve self-analysis by the subject himself.

10 New drugs are usually ...ed on animals before being tried on human beings.

11 Twenty and fifty are ...s of 10.

12 The job of ...ing artificial satellites and of computing their orbits is done by a network of satellite ...ing stations distributed all over the world.

13 A given phenomenon may often be the result, not of one cause only, but of ... interacting factors.

14 When light of a certain wavelength (e.g. ultraviolet) falls on a metal plate—especially an alkaline metal—it causes electrons to be emitted from the plate. This is known as the photo-electric ....

15 There is a noticeable time ..., where the phenomenon is observed from a distance, between a flash of lightning and the sound of the corresponding thunder.

16 Inadequately planned educational systems lead to a great ... of human resources.

(b) Arrange the above completed sentences in pairs so as to show the two different functions of each word from the list, and read them aloud.

Structure Study Revision

TENSES

1 *Present Continuous, Present Perfect, Past Perfect* (see Units 6 and 9):

EXERCISE

In the following passage, put the verbs in brackets into the correct tense:

A chance observation frequently (start) a new field of investigation. For example, Sir Alexander Fleming found that during his absence from his laboratory a mould (grow) on a culture medium containing pathogenic bacteria, and noticed that the bacteria near the mould (stop) growing. This accidental happening and the correct observations and deductions which Fleming (make) from it (lead) to the mass-production of antibiotics which (save) thousands of lives in the last few decades and which (be used) nowadays to fight diseases which up to now (be considered) fatal. In the last few years researchers (discover) other moulds that at present (be applied) successfully against an even wider range of diseases. Research in, and the production of antibiotics is a field that (expand) constantly, and which (succeed) in extending the life-expectancy of millions of people already.

2 *Conditionals* (see Unit 9):

EXERCISE

Complete the following conditional sentences, putting the verbs in italics into the right tense in accordance with the instructions given in brackets in each case:

1 If the current growth rate continues, the world population *multiply* (certain result) six times in the next 100 years.

2 This population increase *cause* (certain result) severe stresses in the economies of developing regions unless widespread and far-reaching measures are taken.

3 If the metal lead (Pb) is cooled to a temperature below 7·2°K (i.e. —265·9°C), it *become* (usual result) a highly efficient conductor of electricity (a superconductor).

4 A body continues in its state of rest or of motion at constant speed in a straight line unless a force acting on it *make* (no obstacle) it change that state.

5 If laboratory instruments *be not* (no obstacle) carefully checked and adjusted before an experiment, misleading results *be obtained* (inevitable result).

6 If politicians *know* (hypothesis only) more about science, they *obtain* (inevitable result) better results.

7 If we put a kilogram of ice at 0°C into 80 kg of water at 1°C, the ice *melt* (normal result) and *cool* (normal result) the whole mixture to 0°C.

8 If average world temperatures *drop* (hypothesis only) by only 5°C a new glacial age *begin* (inevitable result).

9 Unless governments *restrict* (no obstacle) the testing of destructive nuclear devices, widespread contamination of the atmosphere by radioactive fallout may occur.

10 If certain ocean currents *change* (possibility or hypothesis only) their direction slightly, many economically valuable species of fish *be affected* (inevitable result).

ANOMALOUS FINITES (see Unit 8)

EXERCISE

Complete the passage given below, putting the appropriate anomalous finites in the spaces in accordance with the ideas given in brackets:

In planning a series of experiments, the scientific worker ... (moral obligation) be aware of the general nature of the problem under investigation as well as of data from other areas of research which ... (possibility) be related to it. Although in a few cases it ... (possibility) be possible to begin with a completely-organized theory, it ... (moral obligation) be kept in mind that even an imperfect theory ... (physical ability) be very useful, since it ... (possibility) provide a framework which ... (physical ability) later be adjusted to fit the results of the experiments. These preliminary hypotheses ... (moral obligation) be in accordance with the known facts, and this implies that the researcher ... (compulsion) also have a full knowledge of the theoretical background of the problem. After he has analysed the problem he ... (compulsion) present it in as simple a form as possible, since most pieces of work ... (physical ability) be broken up into component parts which ... (physical ability) then be dealt with separately. This procedure ... (possibility) often help to solve the problem more efficiently.

CAUSE-AND-RESULT STRUCTURES (see Unit 9)

EXERCISE

Read the following sentences (which indicate the result of an excess of something) and replace the structure *so* (adj.) *that ... not* (verb) by the structure *too* (adj.) *to* (verb).

e.g. (a) The patient was *so* ill *that* he *could not undergo* the operation, **becomes:**
The patient, was *too* ill *to undergo* the operation.

(b) Our current programmes are *so* extensive *that we cannot cut down* expenses, **becomes:**
Our current programmes are *too* extensive *for us to cut down* expenses.

1 The calculations were so complicated that they could not be done without using a computer.

2 The variations between the experimental group and the control group was so small that it was not significant.

3 The apparatus is so complicated that it cannot be assembled in under 48 hours.

4 The available steels were so weak that they could not be used for the long suspension bridge.

5 The project was so expensive that it could not be continued.

6 A desert is caused when the precipitation (i.e. the quantity of rain or snow which falls) is so small that it cannot offset the loss by evaporation.

7 Many conventional laboratory instruments are so large or heavy that they cannot be used in space research, so special miniaturized versions have been developed.

Discussion and Criticism

1 Apart from the aerospace industry, what other industries can be connected to a substantial extent with defence? Explain how and to what degree.

2 In ll. 60–63 reference is made to government plans requiring the special knowledge of biologists and social scientists. Give details about some of the projects in which these scientists would play an important part.

3 Give details of the various ways in which your own profession plays a part in government action.

4 Discuss the importance of mass media in politics. Which type of medium is likely to exert the most influence, and why?

5 Describe some of the 'far-reaching changes in social patterns' (ll. 51–52) which may follow a process of industrialization.

6 In many countries, a large proportion of the available resources are applied to the development and/or maintenance of destructive apparatus (ll. 11). Find out the relative proportions of the national resources spent on defence, education, agriculture, public health and industrial development, evaluate the reliability and significance of the statistics you obtain, and present the result of your findings in a bar-graph (see Unit 6, Discussion and Criticism section, No. 11).

7 What evidence can you produce, as far as your own country is concerned, to support the statement that the politician 'has in almost every case been educated and trained in a non-scientific or even anti-scientific tradition' (ll. 86–87). What is the evidence *against* this point of view?

8 Would you yourself, as a scientist, be willing to enter politics or exert influence in political affairs at any stage in your career? Give reasons for and against.

9 Make up at least two further projects for this section.

BIBLIOGRAPHY

PIGANIOL (ed.) *Science and Parliament*, Organization for Economic Co-operation and Development.

DON K. PRICE *Government and Science*, Galaxy Books.

C. P. SNOW *The Two Cultures, and A Second Look*, C.U.P. (U.K.) and Mentor Books (USA).

C. P. SNOW *Science and Government*, Mentor Books.

BAGRIT *The Age of Automation*, Mentor Books.

GOLDSMITH AND MACKAY (eds.) *The Science of Science*, Penguin Books.

KELLEY *Professional Public Relations and Political Power*, John Hopkins Paperbacks.

Two periodicals devoted mainly to this and related subjects are:
Impact, published 4 times a year by UNESCO;
Minerva, published 4 times a year by C.S.F. Publications, London.

Supplement

EXTRACTS FROM CURRENT SCIENTIFIC LITERATURE

The extracts appearing in the following short supplement have been selected in accordance with the following criteria:

1 The material included is taken mainly from British and American sources, but also include examples of translations appearing in international journals.
2 In subject, all the main areas of science are represented.
3 They cover the main types (subregisters) of scientific writing, i.e. instructions, descriptions, explanations, abstracts or summaries, hypothesizing and the oral lecture.

The purpose of these readings is to consolidate and extend the knowledge and use of the scientific English which the student has acquired from the Units previously presented.

1

SUBJECT *Biology, Ecology, Radiochemistry*
TYPE OF WRITING *Summary (Abstract)*

Project Title: CYCLING OF CHLORINE-36 LABELED DDT IN A MARSH ECOSYSTEM

Contractor: Ohio State University
Contract: AT(11-1)-1358 Date Activated: 3/1/64
Current Investigators: Tony J. Peterle and Robert L. Meeks

OBJECTIVE

To determine the fate of the insecticide DDT in a marsh ecosystem[1]. Special emphasis is being given to rate and quantity of DDT uptake[2] by aquatic plants and organisms and then to its subsequent translocation[3] within the ecosystem for two growing seasons.

JUSTIFICATION

DDT has been used more than any other insecticide. Its distribution is now almost universal. Scientific publications frequently include reports indicating high residues[4] in organisms far removed from the target species. These accumulations of the insecticide or its metabolites[5] may be harmful. The means of residue accumulation are rarely understood; beneficial organisms are affected long after insecticide application and only by extrapolation can the path of accumulation be estimated. This project is designed to fill the knowledge gap that exists in the time between insecticide application and the occurrence of undesirable effects.

PROCEDURE

On July 7, 1964, 3·9 millicuries of Cl-labeled DDT on inert granules were applied by helicopter to an enclosed four-acre marsh area (Winous Point Club, Port Clinton, Ohio) at the rate of 0·2 pounds DDT per acre. Plant, animal, soil and water samples will be collected and analyzed for DDT using radiochemical techniques at the following intervals after application: four hours, eight hours, one day, three days, one week, two weeks, and then at one-month intervals when selected organisms are available over the 14-month period of field study. Following the compilation of residue data based on the radio-assay[6] results, the distribution and bioaccumulation of the insecticide will be described as it is related to time.

PUBLICATION, THESES AND DISSERTATIONS: None

Offsite Ecological Research of the Division of Biology and Medicine, U.S. Atomic Energy Commission.

[1] *ecosystem:* the relationships existing between plants and animals and their environment in a specific region.
[2] *uptake:* intake.
[3] *translocation:* transfer.
[4] *residues:* small quantities left over from a certain process.
[5] *metabolites:* chemical substances essential to the process of changing food into a form that can be utilized by the living organisms.
[6] *radio-assay:* process of determining the quantities of radioactive elements in a living organism.

Assignments

1 The above extract is an example of a valuable tool for the researcher investigating the literature of his speciality: the abstract. The aim of an abstract is to give a brief account of the main conclusions of a report or any other publication, usually without entering into details. This serves as a quick guide for the investigator faced with a mass of literature and helps him to locate relevant information quickly and efficiently.

(a) Look up in the library of your college or university faculty the titles of the journals specializing in abstracts about your own discipline. Report on these to the rest of the class.

(b) Find and bring to class examples of abstracts about the science you are studying or about a specific problem being dealt with at present in any of the specialized subjects you are studying.

2 Usually an abstract is a summary of 1 to 5% the length of the full article or report. Choose any of the Supplementary Readings that have been dealt with in class and produce an abstract of no more than 5% the length of the original.

2

SUBJECT *Geology, Economics, Engineering*
TYPE OF WRITING *Summary (Report)*

SURVEYING NATURAL RESOURCES

1. This section deals with natural resources such as minerals, water and energy. The United Nations family has considerably increased its activities in the field of natural resources and is intensifying its efforts to secure the application of modern technology to resources development. New approaches are also being explored, such as the preparation of projections combining work on related but different groups of resources.

2. The United Nations Secretariat is organizing training programmes in the field of natural resources for personnel from developing countries, mostly through regional and inter-regional seminars. These seminars are also a useful means of transferring advances in technology to developing countries. They have covered such topics as cartography in relation to development, desalination[1], energy policy and geochemical techniques in mineral exploration.

3 The Secretary-General has prepared a five-year survey programme which is designed to contribute to the development of natural resources by indicating economic and technologically-advanced approaches to the exploration and assessment of these resources. The proposed programmes consists of nine surveys in the fields of mineral resources, water resources, energy and electricity, as follows:

(a) In the field of mineral resources, a survey of off-shore[2] mineral potential in developing areas, a survey of world iron-ore resources, a survey of important non-ferrous metals and a survey of selected mines in developing countries with a view to increasing ore reserves and production through the application of modern technology.

(b) In the field of water resources, a survey of water needs and water resources in potentially water-short[3] developing countries and a survey of the potential for development in international rivers.

(c) In the field of energy and electricity, a survey of potential geothermal energy[4] resources in developing countries, a survey of oil shale[5] resources and a survey of the needs for small-scale power generation in developing countries.

4. Each of the nine proposed surveys would have two objectives: first, to provide significant new information, ideas and approaches on the natural resources potential of each developing country concerned, and secondly, to gather data that would produce a world-wide perspective of the long-term potential availabilities and needs in the selected areas. They would also be useful in preparing and selecting projects for submission to multilateral or bilateral sources of technical and financial aid.

Interim Report on the Development Decade, The United Nations Administrative Committee on Co-ordination, Sub-Committee on the Development Decade, Third Meeting, 14 April 1966.

[1] *desalination:* the process of removing salt from sea-water and in this way make it suitable for human consumption.
[2] *off-shore:* located underneath the sea-bed near the coastline.
[3] *water-short:* suffering a shortage of water.
[4] *geothermal energy:* energy obtained from the internal heat of the earth.
[5] *oil shale:* soft sedimentary rock from which oil can be obtained.

Assignments

1 Imagine that you are participating, as a lecturer, in one of the seminars referred to in ll. 8–10 of the extract. Prepare a short lecture in English on some of the ways in which recent technical advances in your own discipline can help to solve some of the practical problems in developing countries. The rest of the class will act as the personnel being trained, and ask questions.

2 The above extract is an example of a type of scientific writing you may be required to produce in the course of your career. Notice how the subject is presented clearly and systematically by dividing it into sections and subsections. Choose a possible subject for report in your area of interest (e.g. an article you have read, a section of your current textbook, a lecture you have recently heard, a series of experiments you have seen, etc.) and give it a suitable title and sub-titles in English. Then write the corresponding explanation under each sub-title in summary (abstract) form.

3 Prepare in abstract form for the use of an international agency a survey of any one of the natural resources of your own country, keeping in mind the objectives outlined in paragraph 4 (ll. 38–45).

AN EXPERIMENT

PRELIMINARY

The chemistry laboratory is a place where you will learn by observation what the behavior of matter is. Forget preconceived notions about what is supposed to happen in a particular experiment. Follow directions carefully, and see what *actually* 5 *does happen*. Be meticulous (very exact and careful) in recording the true observation even though you 'know' something else should happen. Ask yourself why the particular behavior was observed. Consult your instructor (teacher) if necessary. In this way, you will develop your ability for critical scientific observa-10 tion.

EXPERIMENT 1: DENSITY OF SOLIDS

The density of a substance is defined as its mass per unit volume. The most obvious way to determine the density of a solid is to weigh a sample of the solid and then find out the volume that the sample occupies. In this experiment, you will be 15 supplied with variously shaped pieces of metal. You are asked to determine the density of each specimen and then, by comparison with a table of known densities, to identify the metal in each specimen. As shown in Table E1, density is a characteristic property.

20 *Table E1: Densities of Some Common Metals, g/cc.*

Metal	Density
Aluminium	2·7
Lead	11·4
Magnesium	1·8
Monel metal alloy	8·9
25 Steel (Fe, 1% C)	7·8
Tin	7·3
Wood's metal alloy	9·7
Zinc	7·1

PROCEDURE

Procure (obtain) an unknown specimen from your instructor. 30 Weigh the sample accurately on an analytical balance.

Determine the volume of your specimen by measuring the appropriate dimensions. For example, for a cylindrical sample, measure the diameter and length of the cylinder. Calculate the volume of the sample.

35 Determine the volume of your specimen directly by carefully sliding the specimen into a graduated cylinder containing a known volume of water. Make sure that no air bubbles are trapped. Note the total volume of the water and specimen.

Repeat with another unknown as directed by your instructor.

QUESTIONS

40 1. Which of the two methods of finding the volume of the solid is more precise? Explain.

2. Indicate how each of the following affects your calculated

density: (a) part of the specimen sticks out of the water; (b) an air bubble is trapped under the specimen in the graduated
45 cylinder; (c) alcohol (density, 0·79 g/cc.) is inadvertently substituted for water (density, 1·00 g/cc) in the cylinder.

3. On the basis of the above experiment, devise a method for determining the density of a powdered solid.

4. Given a metal specimen from Table E1 in the shape of a
50 right cone of altitude 3·5 cm with a base of diameter 2·5 cm. If its total weight is 41·82 g, what is the metal?

SIENKO and PLANE *Experimental Chemistry*, McGraw-Hill.

Assignments

1 Answer the questions given in the extract (ll. 40–51) in English.

2 You will notice that in the above extract the writer has used short sentences and frequent repetitions of key-words. This is because he wishes his writing to be both simple and unambiguous. He also divides his instructions into short steps or stages, which are easy to follow. Taking the extract as a model, give clear instructions in English on how to perform an experiment which you yourself have done recently. Add appropriate questions or exercises.

EFFICIENCY IN ENGINEERING OPERATIONS (OPTIMUM CONVERSION)

Unlike the scientist, the engineer is not free to select the problem which interests him; he must solve the problems as they arise, and his solutions must satisfy conflicting requirements. Efficiency costs money, safety adds complexity, performance increases 5 weight. The engineering solution is the optimum solution, the most desirable end result taking into account many factors. It may be the cheapest for a given performance, the most reliable for a given weight, the simplest for a given safety, or the most efficient for a given cost. Engineering is optimizing.

10 To the engineer, efficiency means output divided by input. His job is to secure a maximum output for a given input or to secure a given output with a minimum input. The ratio may be expressed in terms of energy, materials, money, time, or men. Most commonly the denominator is money; in fact, most 15 engineering problems are answered ultimately in dollars and cents. Efficient conversion is accomplished by using efficient methods, devices, and personnel organizations.

The emphasis on efficiency leads to the large, complex operations which are characteristic of engineering. The pro- 20 cessing of the new antibiotics and vaccines in the test-tube stage belongs in the field of biochemistry, but when great quantities must be produced at low cost, it becomes an engineering problem. It is the desire for efficiency and economy that differentiates ceramic engineering from the work of the potter, textile engineer- 25 ing from weaving, and agricultural engineering from farming.

Since output equals input minus losses, the engineer must keep losses and waste to a minimum. One way is to develop uses for products which otherwise would be waste. The work of the chemical engineer in utilizing successively greater fractions of 30 raw materials such as crude oil is well known. Losses due to friction occur in every machine and in every organization. Efficient functioning depends on good design, careful attention to operating difficulties, and lubrication of rough spots, whether they be mechanical or personal.

35 The raw materials with which engineers work seldom are found in useful forms. Engineering of the highest type is required to conceive, design, and achieve the conversion of the energy of a turbulent mountain stream into the powerful torque[1] of an electric motor a hundred miles away. Similarly 40 many engineering operations are required to change the sands of the seashore into the precise lenses which permit us to observe the microscopic amoeba[2] in a drop of water and study the giant nebula in outer space. In a certain sense, the successful engineer is a malcontent[3] always trying to change things for the 45 better.

SMITH, R. J. *Engineering as a Career*, McGraw-Hill.

[1] *torque:* combination of forces producing a rotating or twisting motion.

[2] *amoeba:* single-celled life-form having no definite shape.
[3] *malcontent:* a person who is never satisfied with what is achieved.

Assignments

1 The extract above has been taken from a book whose objective is to introduce engineering as a career to beginning students of that discipline. Consult your specialist teachers or, better still, a practising specialist about the career opportunities, different specializations, duties and responsibilities of your own discipline. Give an oral report in English of your investigation to the rest of the class who should then ask questions.

2 In lines 19–25 it is stated that any problem involving the low-cost production of large quantities of any item is an engineering problem even if the item itself originated in the work of other disciplines. Explain how any given result of (a) medical research, (b) agricultural research, (c) nuclear physics, (d) optical research is likely to need solutions requiring the skills of an engineer.

3 Explain in detail why (a) 'efficiency costs money', (b) 'safety adds complexity', (c) 'performance increases weight' (ll. 3–5). Is this always true? Interview a practising engineer and give an abstract in English of his reply.

4 In lines 5–6 it is stated that the engineering solution to most problems is the 'most desirable end result taking into account many factors'. Does this apply to your own discipline? If so, explain in what way.

5

SUBJECT *Sociology*
TYPE OF WRITING *Explanation*

PREVENTATIVE SOCIOLOGY

Sociologists have tried to reach beyond the 'weighing and measuring' of social problems to an understanding of the social context in which they are generated. Charles Booth's great 'Survey of London Life and Labour' (1889–1903) is a classic example of this kind of work. He tried to discover how it was that an industrial system which created great wealth nevertheless forced nearly one-third of the population of the richest city in the world to live in 'a state of chronic want'. As the title of the Survey implies, to understand poverty 'we need to begin with a true picture of the modern industrial system'.

Although it deals with a very different kind of social problem, the recent study of traffic in towns (*Traffic in Towns: a Study of the Long-term Problems of Traffic in Urban Areas*, 1964) made by Professor Buchanan and his colleagues illustrates the same shift of focus from the problem itself to the system in which it is generated. The argument of the report begins with a simple enough point: 'Vehicles do not move about the roads for mysterious reasons of their own. They move only because people want them to move in connection with the activities which they (the people) are engaged in.' Traffic is therefore a 'function of activities', and because, in towns, activities mainly take place in buildings, traffic in towns is a 'function of buildings'. The implications of this line of reasoning are inescapable: the movement of traffic through the streets of a town cannot be understood without taking into account the activities which take place in the buildings which line them; to get at the roots of the traffic problem we must approach it through the social and economic factors which determine the ways in which activities (and buildings) are arranged in our towns.

These two studies may also be given as examples of the way in which social research can throw light on the connections between 'social problems' and other tendencies in our society which are widely accepted as normal and good. A 'free labour market' was, in Booth's time, regarded as economically desirable—but what if it could be shown that it forces a certain proportion of the labour force to live in poverty? Is traffic congestion an inevitable consequence of giving free rein (i.e. complete liberty) to the economic forces which shape our cities? Other examples may be given. Is the growing economic independence of women one factor in the increase of the divorce rate? Does the crime and delinquency rate necessarily increase if repressive legal sanctions are relaxed? In trying to answer questions of this kind the sociologist is drawing attention to some of the undesirable 'side-effects' of social developments which are otherwise widely considered to be desirable.

KING, M. D. 'Sociology', *The Social Sciences*, ed. D. C. Marsh, Routledge.

Assignments

1 Why is the extract entitled '*Preventative* Sociology'?

2 Explain clearly in English how you would set up a properly-controlled investigation to discover whether there was any connection in your country between poverty and some other social phenomenon (e.g. crime, absenteeism from work, alcoholism, the profession of any particular political or religious views, etc.). What controls would you use?

3 By rearranging the buildings in your city, show how you could reduce the volume of traffic needed for its day-to-day work. Prepare a rough diagram of your solution, and give a clear explanation to the rest of the class in English. The latter should then ask critical questions.

4 Describe some undesirable side-effects of social developments which are otherwise considered to be desirable.

6

SUBJECT *Biology*
TYPE OF WRITING *Instructions/Hypothesizing*

ANTIMICROBIAL SUBSTANCES FROM SEEDS

BACKGROUND

Microorganisms, especially the actinomycetes, produce a variety of antibiotics. As a result, they have been studied very extensively, especially by industrial research laboratories. But although higher plants are known to produce antimicrobial substances, much less work has been done on them.

Most of the work with plants has been concerned with the leaves, stems, roots, fruit or bark. Relatively little effort has been devoted to the seeds. A few investigators have examined seeds, and their data indicate that seeds may be a good source of antimicrobial substances. Some varieties of seeds are resistant to mold growth in germination tests while other seeds are highly susceptible. This is additional evidence that seeds contain substances inhibitory to microorganisms.

SUGGESTED APPROACH

This study could take any one of several directions. If a large variety of seeds is collected, a general survey for the presence of antimicrobial substances could be carried out. This might indicate whether the presence of such substances is widespread in seeds, or whether they are restricted only to certain types of seeds. As described in Reference 6, such a survey can be made using simple techniques, without extracting the antimicrobial substances from the seeds. Perhaps a more intensive study can be made of one particular substance found in one variety of seed. The antimicrobial spectrum of the substance could be determined by testing its effectiveness against various types of bacteria, yeasts and molds.

For a student who is more chemically minded, there is the challenge of extracting, isolating and purifying the antimicrobial agent. To do this, it would be necessary to work out a method of extraction which does not destroy the substance. Try solvents such as chloroform, dilute acid, dilute base, neutral salt solution, acetone, etc. It might be worthwhile to attempt several extraction procedures, and finally select the one which proves to be the most effective.

COMMENTS ON DIFFICULTIES AND EQUIPMENT NEEDS

Solvents such as chloroform or alcohol and acids and bases have antimicrobial properties. You must be careful not to confuse this effect with the substances extracted from seeds. The solvents may be evaporated from the extracts and the residues tested for activity. Acidic or basic solutions may be neutralized before they are tested. Antimicrobial agents are often rather unstable compounds, and treatment with heat or with strong acids and bases should be avoided.

You should be familiar with the basic bacteriological techniques such as the preparation of culture media, sterilization, culture transferring and maintenance, and aseptic technique.

45 Equipment such as petri dishes, bacteriological test-tubes, flasks, a sterilizer or pressure cooker, and a bunsen burner or alcohol burner are required.

The equipment required for the extraction of anti-microbial agents from seeds would include such items as mortar and 50 pestle, beakers, funnels, filters, pipettes and volumetric cylinders. The extraction can be done with simple equipment. Since these extracts will be tested bacteriologically, they should be as free from contaminating microorganisms as possible.

If you decide to study the antimicrobial agent chemically, it 55 would be advisable to have the advice of a chemist in planning the experiments, especially in purifying and concentrating the active substance.

REFERENCES

(All except No. 6 omitted here)

6. FERENCZY, L. 1956. 'Antibacterial substances in seeds', 60 *Nature*, **178**, 639–640.

Research Problems in Biology, Series 4, Doubleday.

Assignments

1 What are the meanings of the following items which appear in Reference 6 (ll. 59–60: 1956; *Nature*, **178**, 639–640)? For the benefit, say, of an English-speaking investigator who has arrived in your city, explain specifically the different places he could go to in order to find books and periodical literature in your own speciality.

2 Imagine you are contributing to an international textbook which (like the UNESCO *Source-book for Science Teaching*) is to be used in schools and colleges which do not have adequate scientific equipment: prepare *clear instructions* in English on:

(a) how to make a crude extract of a possible antimicrobial substance from given seeds.

(b) how to test the effect of the substance extracted using common articles such as tins and jars instead of expensive laboratory equipment.

3 Describe clearly (so that a non-scientist would be able to understand):

(a) any of the pieces of apparatus mentioned in ll. 45–50,

(b) any of the processes listed in ll. 43–44.

4 The above reading passage is taken from a series of books whose object is to suggest real, important, but relatively simple projects for investigation which can be undertaken by comparatively early-stage students of biology, so that they 'learn by doing'. After consulting your teachers if necessary, try to outline a similar type of real (perhaps local) investigation suitable for students of your own discipline.

7

SUBJECT *Medicine, Agriculture, Biochemistry*
TYPE OF WRITING *Discussion*

THE RESISTANCE OF INSECT PESTS TO INSECTICIDES

. . . The ultimate type of resistance is that in which the insect changes its normal physiology[1] so that it is no longer sensitive to the insecticide. A change of this kind seems to be the explanation of the type of resistance involving a large number of chlorinated compounds like dieldrin. The mode (method) of action of these compounds, however, is quite obscure, so that at present it is scarcely possible to discover how insects become immune[2] to them.

Research in the past fifteen years has revealed a great deal about the nature of resistance, but in no single case have we[3] been able to overcome it completely. In other words, when resistance has developed to a particular insecticide, no means have been found to restore permanently the former effectiveness of that insecticide.

Considering the present situation, it may cause surprise, in view of the large number of reports of resistance from so many important species all over the world, that the impact on insect control programmes is not more drastic (severe). There are two reasons for this. Firstly, many instances of resistance are more or less localized. For example, dieldrin resistance in the major African malaria vector[4], *Anopheles gambiae*, is confined to the west of Africa, though the mosquito occurs in East and South Africa and is equally attacked by insecticides in those regions. One may begin to hope that the genetical potential for developing resistance is lacking in some natural populations of pest insects. Secondly, only a limited number of species show resistance to the two groups of chlorinated insecticides. Until this double resistance develops, it is possible to use either one or the other of these two classes of insecticide and still maintain effective control. Unfortunately, however, the instances of double resistance are growing. By 1960, twenty species of public health importance had developed resistance to both groups of chlorinated insecticides. In addition, four species had developed resistance to organo-phosphorus compounds as well—in other words, treble resistance. It must, then, be concluded that resistance is likely to become a more severe problem in the future than it is at present.

Naturally, a great deal of thought has been given to possible ways of preventing the emergence of resistance. One suggestion has been the use of mixtures of two different types of insecticide, with the idea that one of them should eliminate the individuals resistant to the other. This principle has been found useful in preventing resistance to antibiotics in bacteria. Unfortunately, the few practical trials have not been encouraging, for the mixtures have merely developed a double resistance to the two insecticides employed (used).

In brief[5], there is as yet no known way of obtaining the benefits of the new insecticides without some risk of provoking

(stimulating) resistance. For this reason, it would seem unwise
50 to use insecticides regularly, on a very large scale, unless there is some vital object to be attained. In such cases, the use of insecticides should be combined with other measures, for examples (as regards insect-borne disease) strong efforts to improve general hygiene[6].

BUSVINE, J. R., 'The Challenge of Insecticide Resistance', *Penguin Science Survey B, 1966*, Penguin Books Ltd.

[1] *physiology:* structure and manner of functioning.
[2] *immune:* unaffected by, resistant to.
[3] *in no single case have we . . .:* an emphatic way of saying 'we have not, in a single case . . .'
[4] *vector:* in medicine, an organism which carries or transmits the causal agents of disease to man, animals or crops.
[5] *in brief:* to explain in a short space, to summarize.
[6] *hygiene:* practical procedures for healthy living.

Assignments

1 The problem outlined in the above extract is a good example of a field of research in which a solution depends on the joint efforts of investigators from several different disciplines, such as chemistry, entomology, genetics and parasitology. Discuss this point and explain briefly in English how you think representatives from each of these disciplines would tackle the problem.

2 Find out about, and explain in simple English to the rest of the class, the phenomenon of resistance in any of the following forms:

biology: resistance of living organisms to extremes of temperature;
resistance of living organisms to disease;
resistance of living organisms to radiation.
(NOTE: Specific examples may be dealt with where convenient.)

psychology: resistance of human beings to suggestibility (see also the extract 'The Pressure to Conform').

sociology: resistance of human groups or institutions to change or new ideas.

engineering: resistance of materials.

3 In what other ways can insect-carried disease in medicine and agriculture be controlled, apart from the use of insecticides?

4 Describe clearly, for the benefit of an English-speaking visitor, the nature and extent of insect-borne diseases in medicine or agriculture in your own country, and the ways in which they are being controlled.

8

SUBJECT *Medicine, Biochemistry*
TYPE OF WRITING *Explanation*

CHELATION IN MEDICINE

Finding new drugs for the treatment of human diseases is still largely fortuitous.[1] Most of the drugs in current use were discovered by accident or by trial and error, and the cases in which a clear connection has been found between a drug's action in the body and its chemical and physical properties are few. One class of drugs for which such a connection has been established, however, is the group known as the chelating agents. These substances are characterized by their ability to seize and 'sequester'[2] metal atoms. Since their various actions as drugs are apparently based at least in part on this property, it offers a promising foundation for the development of a rational pharmacology. The promise is being vigorously explored in laboratories all over the world.

Chelation is a common chemical phenomenon and is associated with many familiar substances. Among the well-known natural chelates are hemoglobin (containing iron), chlorophyll (containing magnesium) and vitamin B-12 (containing cobalt). Among the common substances that can act as chelating agents are citric acid, aspirin and a host of other compounds, natural and synthetic. The phenomenon of chelation was recognized many years ago and put to use in various applications, although it was then poorly understood. In 1935 a German chemist, F. Munz, patented a compound called ethylenediaminetetra-acetic acid (EDTA) that had a remarkable affinity for calcium. It soon found commercial use in the textile industry as an agent for preventing the precipitation of calcium from the water used in the manufacture of fibers.

It was only a little more than a dozen[3] years ago, however, that chemists in general became aware of the fundamental importance of chelation and began to devote intensive study to its possible commercial and medical applications. The medical potential of chelating agents was demonstrated in 1951 when EDTA saved the life of a child suffering from lead poisoning. Since then the investigation of the medical uses of chelating agents has produced a voluminous literature, and chelate drugs have been developed for the treatment of a wide range of diseases.

A chelating agent is a molecule that is capable of seizing and holding a metal ion in a clawlike grip (the term comes from the Greek word *chele*, meaning 'claw'). Like a claw, the structure forms a ring in which the ion is held as if by a pair of pincers. The pincers consist of 'ligand'[4] atoms (usually nitrogen, oxygen or sulfur), each of which donates two electrons to form a 'co-ordinate' bond with the ion. In most cases the metal ion can be grasped by more than one molecule, so that the ion is held in a set of rings. Each ring is usually composed of five or six members consisting of single atoms or groups of atoms.

The medical importance of chelating agents hinges on the fact that metals play many critical roles in the life of living organisms. In the human body, metabolism depends not only on

sodium, magnesium, potassium and calcium, but also to a con
siderable degree on trace[5] amounts of iron, cobalt, copper
zinc, manganese and molybdenum. On the other hand, certain
other metals, even in minuscule[5] amounts, are highly toxic to
the body. It is apparent, therefore, that chelate drugs with
appropriate properties could play several different therapeuti
roles. Various chelating agents might be designed (a) to seek ou
toxic metals and bind them in compounds that will be excreted
(b) to deliver essential trace metals to tissues or substances tha
require them, and (c) to inactivate bacteria and viruses by de
priving them of the metals they need for their metabolism or by
delivering metals to them that are harmful. All three of these
possibilities have been realized.

SCHUBERT. J. in *The Scientific American*, May 1966.

[1] *fortuitous:* happening by chance or accident.
[2] *sequester:* to isolate. In this particular context, to render inactiv or inert.
[3] *a dozen:* twelve.
[4] *ligand:* those that bind or tie.
[5] *trace: minuscule*: very small, minute.

Assignments

1 The word 'chelation' comes from a Greek 'root' (as ex plained in ll. 39–40) plus the suffix '-tion'. A great deal of th special vocabulary used in science is in fact made up from Gree and Latin roots, and it is worth while learning the commonest c these.

Make a list of the words belonging to the science you yoursel are studying which are made up in this way, and find out th meaning of the Greek and Latin root(s) in each case. Arrang them in alphabetical order for future reference.

2 This extract explains what is often called a 'growing poin in the particular sciences concerned (medicine and biochemistr in this case). A 'growing point' is a comparatively new an fruitful development in which the essential background wor has been established, but where the implications and applica tions still remain to be fully exploited. By talking to you specialist teachers, find out some 'growing points' in the scienc you yourself are studying, and give a clear talk on ONE of thes in English, to the rest of the class. The rest of the class shoul ask questions or give comments, also in English.

OPERATIONAL RESEARCH AND SOCIAL CHANGE

In war the desire for survival, which demanded the most effective use of all the resources of the country, and thus many social changes, brought about the birth of operational research. Examples of the use of O.R. during the war were the develop- 5 ment of better convoy systems, better attack techniques, etc. In terms of social change this meant that it was now accepted that activities previously regarded as the sole province (speciality) of a traditional specialist, the professional soldier for instance, could be usefully commented on by specialists from other fields, 10 for example scientists.

After the war operational research moved into industry. Its development coincided with a desire not now just to survive but to set up a new society which would be better than that of pre-war years. Industry everywhere faced rising labour costs and 15 more complex processes, both arising from the determination of human beings to achieve a better standard of living. This desire stimulated operational research, and at the same time the tools—large-scale electronic computers and sensitive electronic-control devices—were ready to take advantage of the development.

20 What changes, that can be described as social, flow out of all this? To start with, mathematicians become valued members of the industrial community, and are no longer destined merely to teach or lecture. The manager now has to include in his bag of tricks[1], queueing theory, simulation, network analysis, linear 25 programming, and other mathematical techniques all devised primarily by O.R. specialists. He has to accept that areas in which he might feel happy and secure about using his own experience, are now increasingly the province of the O.R. analyst, and depend on equipment operated by programmers 30 talking a strange language of their own. On the shop floor[2], again, control equipment designed in conjunction with (together with) an engineer knowledgeable in some of the techniques of O.R. is taking decisions away from the operator. Thus to both management and labour O.R. could mean the removal of 35 decisions and a move towards boredom and frustration.

A recent official report estimated that about 20,000 people are recruited to management in the United Kingdom each year. Only a small percentage of them will have the educational background adequate to cope with (deal with) the effective use of 40 these new techniques, and we have too, the general problem of a low average level of education which prevents effective communication on this subject. It is not easy, for example, to explain the processes involved in a computerized control system if the basic algebraic symbols are without meaning.

45 Let us consider a little further one of changes which can arise from O.R. applications in industry. I have already mentioned that these techniques can be used even when they are not well understood by the lower levels in management, and that this

will generally be associated with the centralization of decision.
50 This largely arises from the scarcity (shortage) of the knowledge required to carry out operational research and operate the necessary computing equipment; it appears economical to use this scarce resource best by concentrating it at one point. The decision processes are thus completely removed from the whole
55 level of middle management personnel and placed at some central point, possibly hundreds of miles away. But what are the alternatives? One cannot suddenly train a large number of middle management personnel in O.R. techniques or suddenly give them confidence in the value of scientific method. Thus we
60 may be faced with a society having two basic groups in industrial organization: one consisting of those who understand, through their education and training, the complicated techniques necessary to control the complex industrial and organizational environment that has developed; the other consisting of
65 those who carry out the decisions of this organizing group. We have to decide if we want this type of society; and if not, what sort of society we *do*[3] desire.

DUNN, H. D. broadcast talk reprinted in *The Listener*, 3 February, 1966.

[1] *bag of tricks:* a humorous way of referring to specialized knowledge.
[2] *on the shop floor:* in the workshop.
[3] *do* (structural): an example of the use of 'do' to make the meaning of the verb that follows ('desire' in this case) stronger and more emphatic.

Assignments

1 Operational research is an example of a field of research which was strongly stimulated by war. Give examples of other sciences that have been stimulated in this way.

2 In line 18 reference is made to 'sensitive electronic-control devices'. Consult your specialized teachers and give an oral report in English to the class describing the types of similar devices that are used in your own discipline and giving a general idea of how they work.

3 The extract above deals with the social effects of a newly-developed technique. Describe, for the benefit of an English-speaking visitor, the social changes that have been brought about by the application of science in your country in the last ten years.

10

SUBJECT *Agriculture, Biology*
TYPE OF WRITING *Explanation*

PLANT BREEDING FOR THE DEVELOPING NATIONS

The application of the science of genetics to plant breeding occupies a strategic place in the enhancement (increase) of crop productivity. Upon its success depends the effectiveness of many other efforts to provide adequate food supplies for direct consumption by man, feed for his animals and agricultural raw materials for his industries. Bringing more water to the land, improving the soil, providing plant nutrients, teaching the farmer better cultural practices, giving him more efficient tools, etc., cannot yield maximum results unless the plants under cultivation are able to respond fully to the improved environment and practices.

The application of the principles of plant genetics and of the technical methods used in plant breeding to the needs of the developing countries starts off with the advantages of there being adequate labour and, usually, land at the disposal of the breeder. Lack of equipment may not be serious (i.e. a detrimental factor)—many of the major advances in plant breeding have been made in the field, including the basic work of Mendel.

Plant breeders' main aims are the same throughout the world. Apart from evolving varieties with higher yield potentials, the usual main objectives are greater cultural reliability, greater resistance to diseases and pests, accommodation (adaptation) to the special requirements of different types of cultural practices, and improvement of quality, both nutritional and industrial. Most of the scientific methods developed are equally applicable everywhere. In fact, plant breeding work is already in progress in most developing countries, and has been begun in some instances by local workers on their own initiative, or with the assistance of international bodies such as FAO.

When starting breeding work to improve crop production, particularly of food crops, developing countries should follow simple and inexpensive methods which, in the great majority of cases, are economic and efficient. The first step should be the examination of the available seed sources. These include such indigenous materials as local varieties in farmers' fields and natural stands. The plant breeder is also able at such an examination to assess the characteristics of the crops which are most in need of improvement. The organized distribution of seed of the best varieties can, within a very short time, make an important impact on the country's crop production.

The second potential source of 'ready-made' varieties are introductions from other parts of the world. Agro-climatic analogues (analogies) exist between various countries, and varieties adapted in one country may, in general, be expected to be adaptable to another having similar ecological and climatological conditions.

Selections made from the indigenous and imported materials will usually provide suitable populations for the next step, i.e.

the more formal breeding work, which should be started on the
50 broadest possible heterogeneous (diverse) base. When relevant,
it should be noted that large collections of germ plasm of wheat,
oats, barley, maize, rice, soya beans, lucerne and other crops are
available in most of the advanced countries.

Ultimate progress will depend on developing populations
55 which have desirable combinations of genes and gene frequencies. To make sure of the presence of these genes, hybridization of carefully selected parents will be necessary. Frequency
of genes can be increased by various mating systems, the
technical details of which are set out in Conference Paper
60 C/443.

Once suitable populations are available, success will depend
on effectiveness of selection. To ensure this, the differential
effects of environment among the individuals must be reduced
to a minimum. This can be done by:

65 (a) special measures aimed at ensuring uniform environment
for all plants in the population;

(b) the use of experimental designs which will reveal differences, control and measure error, and provide a valid
estimate of the significance of the differences; and

70 (c) the use of selection units appropriate to the characters for
which the selections have been made.

'Agriculture', *Report on the United Nations Conference on the Application of Science and Technology for the Benefit of the Less Developed Areas*, **III**, U.N.O., 1963.

Assignments

1 Imagine a foreign agronomist is visiting your district. As you know some English, you have been asked to explain to him the main features of the agriculture of the zone. Do this as clearly as you can.

2 Give a short lecture in English on EITHER: one of the factors affecting agriculture or stock-raising in your country (e.g. climate; types of soils; water supplies; farming practices; system of land ownership; types of crops or animals reared; marketing arrangements, etc.), OR: some of the problems affecting agriculture in your country.

3 Find out about *germ plasm* l. 51—what it is, how it is prepared and how it is used—and explain it in English to the rest of the class. Do the same for *mating systems* (l. 58) and the *design of experiments* (l. 67).

4 Give a talk describing the work of Gregor Mendel.

WHAT IS PSYCHOLOGY?

References to psychology in the daily press and in popular periodicals are now numerous, but the variety of ideas as to the nature of psychology is correspondingly extensive.

The existence of so many varied conceptions of the nature of
5 psychology is no doubt related to the many aspects of psychological work. The psychiatrist dealing with 'mental' breakdown, the educator moulding human development, the vocational counsellor advising on the choice of jobs, the social scientist studying the prevention of crime, the personnel manager smooth-
10 ing human relations in industry, the industrial psychologist streamlining industrial processes to suit the nature of human capacities: all these are concerned with psychology. Human behaviour is complex and varied, and the science which studies it must have many aspects. It may be useful for us to consider
15 samples of psychological work in various areas.

CASE I—A SIMPLE INDUSTRIAL PROBLEM

A large London catering firm became concerned about the excessive number of breakages by its employees. It therefore decided to impose a penalty to reduce carelessness. Strange to say, the system of fines led to an increase in breakages. The
20 management decided that the problem was more complex than they had thought, so they called in a psychologist to investigate and recommend appropriate remedies. The psychologist first posed the question as to when breakages occurred. He kept a record of the breakages occurring during half-hour periods over
25 a number of days, and it soon became evident that most accidents occurred during the rush periods when the girls were worried by their inability to cope with the number of orders. It was now obvious why the system of fines had only made matters worse. They added to the anxiety of the already over-strained girls and
30 simply made them more nervous.

CASE II—READING DIALS

The handling of modern planes (aircraft) places a severe strain on the pilot who must deal with many things in rapid succession. He has to keep watch on a number of dials which give him vital information about his speed, altitude, etc. It is essential
35 that these dials should be easily read and not subject to error. What form should they take? They may have vertical scales or horizontal scales, they may be semicircular or completely round: the whole scale may be visible or only part through a small 'window'. Only careful trials with a number of observers can
40 establish which is the preferred form. An investigator (Sleight) conducted some research to discover the best form for such dials and found that the window type is the best. This seems reasonable since only a part of the scale appears in the window, and so there is less effort required to read the precise point on the
45 scale, but the psychologist has learned never to trust reasoning of this kind. Reasoning may suggest the answer, but we must always try it out.

CASE III—LEARNING

If you have to learn a skill or commit something to memory, the question arises as to whether you should complete the job
50 in one sitting, or whether it is better to spread the learning over a number of periods. A number of experiments have been carried out to decide this matter and, although results vary according to specific circumstances, the general trend is quite definite. In one of the investigations, two similar groups of
55 subjects were set to memorize material. The first group read through the material sixteen times in one day. The other group also read through the material sixteen times, but spread the readings over sixteen days at the rate of one per day. Each group was tested a fortnight (14 days) after completion of the learning
60 to see how much had been retained. The results showed a startling difference. The first group remembered 9% of the material while the second group remembered 79% despite the fact that there was so much opportunity to forget during the 16-day period of learning.

ADCOCK, C. J. *Fundamentals of Psychology*, Penguin Books Ltd.

Assignments

1 From the examples mentioned in the passage (ll. 6–12) and the cases described, try to formulate a satisfactory and reasonably complete definition of Psychology.

2 Describe any examples known to you in which psychology has been applied in any of the following fields: industry; medicine; education; crime; government.

3 Imagine you were the investigator involved in Case II. Prepare a short lecture (in English) to be given before an international audience, explaining the problem and giving details of the way you solved it.

4 Case I depended for its solution mainly upon accurate measurement; Case II depended upon designing an adequate experimental set-up. In your own local surroundings, can you think of any simple problem which could be solved by one, or both, of these methods? Describe clearly one of these real-life problems, and indicate how you would proceed to tackle it.

5 As you know, a great deal of work has been done on the psychology of learning. Nevertheless, the methods employed by the majority of university and college students are notable for their inefficiency. Can you suggest how to plan and undertake a scientific investigation designed to improve their study methods?

THE SCOPE OF GEOLOGY

The world we live in presents an endless variety of fascinating problems which excite our wonder and curiosity. The scientific worker attempts to formulate these problems in accurate terms and to solve them in the light of[1] all the relevant facts that can 5 be collected by observation and experiment. Such questions as What? How? Where? and When? challenge him to find the clues that may suggest possible replies. Confronted by the many problems presented by, let us say[2], an active volcano, we may ask: What are the lavas made of? How does the volcano work 10 and how is the heat generated? Where do the lavas and gases come from? When did the volcano first begin to erupt and when is it likely to erupt again?

Here and in all such queries What? refers to the stuff[3] things are made of, and an answer can be given in terms of 15 chemical compounds and elements. The question How? refers to processes—the way things are made or happen or change. The ancients[4] regarded natural processes as manifestations of power by irresponsible gods; today we think of them as manifestations of energy acting on or through matter. Volcanic 20 eruptions and earthquakes no longer reflect the erratic behaviour of the gods of the underworld: they arise from the action of the earth's internal heat on and through the surrounding crust. The source of the energy lies in the material of the inner earth. In many directions, of course, our knowledge is 25 still incomplete: only the first of the questions we have asked about volcanoes, for example, can as yet be satisfactorily answered. The point is not that we now pretend to understand everything, but that we have faith in the orderliness of natural processes. As a result of two or three centuries of scientific 30 investigation we have come to believe that Nature is understandable in the sense that when we ask her questions by way of appropriate observations and experiment, she will answer truly and reward us with discoveries that endure.

Modern geology has for its aim the deciphering of the whole 35 evolution of the earth from the time of the earliest records that can be recognized in the rocks to the present day. So ambitious a programme requires much subdivision of effort, and in practice it is convenient to divide the subject into a number of branches. The key words of the three main branches are 40 the materials of the earth's rocky framework (mineralogy and petrology); the geological processes or machinery of the earth, by means of which changes of all kinds are brought about (physical geology); and finally, the succession of these changes, or the history of the earth (historical geology).

45 Geology is by no means[5] without practical importance in relation to the needs and industries of mankind. Thousands of geologists are actively engaged in locating and exploring the mineral resources of the earth. The whole world is being searched for coal and oil and for the ores of useful metals.

50 Geologists are also directly concerned with the vital subject of water supply. Many engineering projects, such as tunnels, canals, docks and reservoirs, call for geological advice in the selection of sites and materials. In these and in many other ways, geology is applied to the service of mankind.

55 Although geology has its own laboratory methods for studying minerals, rocks and fossils, it is essentially an open-air science. It attracts its followers to mountains and waterfalls, glaciers and volcanoes, beaches and coral reefs in search for information about the earth and her often puzzling behaviour. 60 Wherever rocks are to be seen in cliffs and quarries, their arrangement and sequence can be observed and their story deciphered. With his hammer and maps the geologist in the field leads a healthy and exhilarating life. His powers of observation become sharpened, his love of Nature is deepened, and the 65 thrill of discovery is always at hand[6].

HOLMES, A., *Principles of Physical Geology*, Nelson, (U.K.), and Ronald Press, (U.S.A.)

[1] *in the light of:* taking into account, considering.
[2] *let us say:* for example.
[3] *stuff:* material.
[4] *ancients:* people who lived many centuries ago.
[5] *by no means:* certainly not (an emphatic negative).
[6] *at hand:* near.

Assignments

1 Describe in English some of the questions relating to the science you are studying that cannot yet be satisfactorily answered.

2 The extract given above is taken from the introductory part of a standard textbook on geology. Its aim is to give a general survey of the scope of the subject, and at the same time to interest the student in his future career. Try to prepare a similar introduction to your own subject, or to the aspect of it which interests you most.

3 In most sciences there are two different sides to investigation, i.e. laboratory work and field, or practical work. Describe clearly an example of each from the discipline you yourself are studying.

THE PRESSURE TO CONFORM

Suppose that you saw somebody being shown a pair of cards. On one of them there is a line, and on the other, three lines. Of these three, one is obviously longer than the line on the first card, one is shorter, and one the same length. The person to whom these cards are being shown is asked to point to the line on the second card which is the same length as the one on the first. To your surprise, he makes one of the obviously wrong choices. You might suppose that he (or she) perhaps suffers from distorted vision, or is insane. But you might be wrong: you might be observing a sane, ordinary citizen, just like yourself. Because, by fairly simple processes, sane and ordinary citizens can be induced to deny the plain evidence of their senses—not always, but often. In recent years psychologists have carried out some exceedingly interesting experiments in which this sort of thing is done.

The general procedure is this: Someone is asked to join a group who are helping to study the discrimination of length. The victim[1], having agreed to this seemingly innocent request, goes to a room where a number of people—about half a dozen—and the experimenter, are seated. Unknown to the victim, none of the other people in the room is a volunteer, like himself; they are all in the league with (i.e. collaborating with) the experimenter. A pair of cards, like those I have described, is produced; and everyone in turn is asked which of the three lines on the second card is equal to the line on the first. They all, without hesitation, pick[2]—as they have been told to pick—the same *wrong* line. Last of all comes the turn of the volunteer. In many cases, faced with this unanimity, the volunteer denies the plain evidence of his senses, and agrees.

Stanley Milgram of Harvard used sounds instead of lines, and the subjects were merely asked to state which of two successive sounds lasted longer. The volunteer would come into a room where there was a row of five cubicles with their doors shut and coats hanging outside, and one open cubicle for him. He would[3] sit in it and put on the earphones provided. He would then hear the occupants of the other cubicles tested in turn, and each would give the wrong answer. But the other cubicles were, in fact, empty, and what he heard were tape-recordings manipulated by the experimenter. Milgram conducted a series of experiments in this way, in which he varied the pressure put upon the subjects, and clearly showed that, faced with the unanimous opinion of the group they were in, people could be made to deny the obvious facts of the case in up to 75 per cent of the trials. I find this more than a trifle[4] alarming—and very thought-provoking.

You may reply that there is no cause for alarm, because in real situations the total unanimity of a group is rare. The more usual case concerns the effects of what we might call a 'pressure group'.

This has been examined, at least partially, by W. M. and H. H.
50 Kassarjian. They used the 'group in a room' and 'lines on cards' situation, and made things much easier for their volunteers. In the first place, the genuine volunteers were in a majority: 20 out of 30. Secondly, the volunteers never had to make their selections aloud, but always enjoyed the anonymity of paper and pencil.
55 The experimenter explained that some people would be asked to declare their choices publicly, and then asked only his 'primed'[5] collaborators. Thus each volunteer heard the views of only a third of the group he was in. Nevertheless, a substantial[6] distortion was still produced: almost, though not
60 quite, as large as in the conditions we looked at first. So there is only small comfort here.

HAMMERTON, M. broadcast talk reprinted in *The Listener*, 18 October, 1962.

[1] *victim:* Here, a humorous way of referring to the subject of the experiment.
[2] *to pick:* to choose.
[3] *would:* In this context, *would* (used for all persons) is used to express actions in the past which were repeated regularly and thus constituted a habitual procedure or routine.
Note the different use of *would* in l. 55, where it is the past tense of *will*.
[4] *more than a trifle:* very.
[5] *primed:* previously prepared by the experimenter.
[6] *substantial:* considerable.

Assignments

1 Imagine that you are the experimenter in the experiments outlined in the extract, and that the other members of your class are 'subjects' and 'collaborators' respectively. For each experiment, explain clearly in English.

(a) what the 'collaborators' have to do,
(b) what the 'subjects' have to do.

Both 'collaborators' and 'subjects' may ask further questions (in English) to make sure that they understand the instructions given.

2 In this sort of experiment, the degree of conformity (i.e. percentage of conformers) probably varies according to the psychological 'pressure' exerted by the experimenter. Can you design some experiments which could establish this point? If so, explain the procedure clearly in English.

3 Explain some of the dangers to society revealed by the experiments outlined in the extract.

4 It is possible that individuals vary in their susceptibility or resistance to suggestibility. How could you find out whether there are certain factors which

(a) predispose people to conform,
(b) enable people to resist false conformity.

and if so, what these factors are?

NOTE: This is a matter of great practical importance in view of Question (3) above.

WATER SUPPLIES—A GROWING PROBLEM

Our need for water is constantly increasing. There is an automatic increase due to population growth, while the overall improvement of living standards, the fight against hunger through the irrigation of more land for food growing, and the creation and expansion of new industries, all foretell the need for even greater water supplies throughout the world. Though it is difficult to calculate the exact amount, it is safe to say that in 20 years' time the demand for water will be roughly double. Faced with such a situation it is obvious that we should search as widely as possible and with every available means for sources of fresh water that seem to be the least costly. But where do these sources exist? Only a sustained and co-ordinated programme of scientific observation and research in hydrology will tell us the answer. This is the purpose of the International Hydrological Decade, 1965–1975.

Underground water reserves are much larger than those on the surface, but as they are unseen we tend to underestimate them. It is vitally important that we make use of these underground reserves, but never haphazardly.[1] For example, where does the water come from which we find in one or another of the underground water-bearing layers ('aquifers')? How does it move? How is it renewed? And if this water is used, what effect will it have on the discharge and future level of the water table? What are the laws of hydrogeology? Despite[2] the immense progress of recent years, all these questions have still not been fully answered.

A similar need for scientific research exists in the branch of hydrology that deals with the quality of water. In nature, there is no water like the pure water defined by chemists, made up of only hydrogen and oxygen. River water, ground-water[3], and even rainwater always contain other dissolved or suspended elements, and these, even when present in small quantities, play an important role. In the case of irrigation farming, for instance, every drop of water brings with it a little salt: the water evaporates, but the salt remains and gradually poisons the soil and plants. In general, we now know how to remedy this problem of salinity with the help of leaching and drainage. But many questions remain unanswered regarding the effect of irrigation and drainage on the quality of ground-water, and the possibility of maintaining the ground-water level below the zone of the plant roots while bringing to the surface the water necessary for irrigation.

What happens exactly in this thin layer of soil which preserves the moisture necessary to plant life? What form—liquid or vapour—does the water take in this zone? What forces act on the water, depending on the kind of soil present? How long will this life-giving moisture last?

Evaporation from the soil and transpiration from vegetation are responsible for the direct return to the atmosphere of more

50 than half the water which falls on the land. How exactly do these phenomena, which represent an enormous loss of resources, occur? What part does a forest play in the water balance-sheet[4] of a given area? Does it act merely as a water-consuming mechanism operating through the absorption and transpiration 55 of the trees—thereby reducing the quantity of runoff which reaches the rivers—or, on the contrary, does it result in a slow seepage into the earth which can later be recovered in the form of ground-water, while at the same time preventing erosion?

These are the kinds of problems which still have to be re-60 solved: the answers will only be found through a vast programme of scientific research.

BATISSE, M., *Courier*, UNESCO, July-August 1964.

[1] *haphazardly:* at random; without having sufficient knowledge.
[2] *despite:* in spite of; notwithstanding.
[3] *ground-water:* sub-surface water forming a saturated zone between ground level and the topmost layer of impermeable rock.
[4] *balance-sheet:* a statement of gains and losses.

Assignments

1 This extract is an example of a growing type of literature with which scientists all over the world have to deal, i.e. translations from a foreign language (in this case, French) into English. These may be done by individuals—in which case they may be very imperfect and may present actual misinformation for the inattentive reader—or by commercial firms or the specialized Translation Sections of the main international organizations, as in the present extract.

Imagine that a translation, for an international journal, is required of a short summary describing EITHER some of the scientific work being done at the institution you are studying at, OR a national problem connected with the science you are studying.

Ask one of your specialist teachers to write about 100 words in the vernacular on one of the above alternatives (or do it yourself if you have enough knowledge), and then produce the translation required.

2 You will notice that the passage contains a large number of questions:

(a) Imagine that you are attending an international seminar for science students on 'Planning Research Projects': you have been asked to give suggestions on how to plan an investigation into any one of the questions raised in the passage. Give these in English.
(b) Formulate similar questions 'still not fully answered' regarding the discipline you are studying yourself.

15

SUBJECT *Engineering, Mathematics, General*
TYPE OF WRITING *Explanation*

DIGITAL COMPUTERS AND THEIR USES

In the digital computer the numbers to be manipulated are represented by sequences of digits which are first recorded in suitable code—usually the binary code—, are then converted into positive and negative electrical impulses, and stored in 5 electrical or magnetic registers which serve basically the same purpose as the counting wheels in a desk calculating machine.

The technique of making the computer carry out a particular calculation is known as 'programming', which involves first breaking the calculation down into a sequence of arithmetic 10 operations, and then preparing a series of instructions which cause the computer to carry out the required operations on the stored information in the correct order. It is now possible to add or subtract two large numbers in one to two microseconds, and to multiply or divide them in ten to twenty microseconds, so that 15 a computer can perform as much arithmetic in a quarter of an hour as an efficient clerk with pencil and paper might reasonably hope to achieve in a lifetime.

There are many situations in which this ability to handle and to analyse large quantities of arithmetic data according to 20 instructions is of great value. Some examples are fields of scientific investigation such as crystallography, atomic physics and astronomy, where masses of experimental data are involved and complex theoretical concepts need to be tested against them; in engineering design where the design parameters[1], of which 25 there are many, can be varied systematically and their effects studied and optimized; and for the storage of reference data in libraries and insurance offices in such a way as to afford ready access to particular references on request.

A particularly important application of the digital computer 30 in simplified form is as a component in the control equipment of manufacturing processes—as the nerve centre which accumulates and analyses data recording the operating conditions and performances of the plant, and sends out instructions, when appropriate, for their modification. This is one aspect of what is called 35 'automation'—the replacement of human control by instrumental control. The completely automatic factory is no longer a fantasy.[2] What is restraining its realization are the difficulties in handling the severe economic considerations and the complex human problems involved.

40 During a recent conference on 'Computable Models in Decision-taking', the chairman said:

> 'It is significant that despite (in spite of) the rapid advance of science since the seventeenth century, it made no impact on the problems of prediction until the advent of the digital 45 computer, and it is only thanks to (owing to) the powerful new computers that worth-while prediction in human affairs has been possible at all.'

He went on to say that accurate prediction, and therefore decision-taking, is only possible in autonomous systems for

50 which the laws of behaviour are known. This is not, of course, the case in human affairs, and the most that can be yet done is to seek for mathematical models which describe, however imperfectly as yet[3], the presumed behaviour of the system or situation under investigation, and then to study systematically,
55 with the aid of the computer, the consequences which arise from the variation of the parameters incorporated in the model. It was fascinating to hear of some of the problems falling within the sphere of human behaviour to which the computer is now being applied experimentally. One was that of forecasting the sale of
60 particular kinds of fabric, with a view to ensuring that out-of-stock situations do not develop; another, predicting the future demand for various categories of steel products, as a basis for judging the necessary provision of manufacturing capacity; and a third, an attempt to establish a model of the economic system
65 of the country, as a means of predicting its future pattern of development.

The point that I am anxious to make is that the search for models of this kind, the study of their behaviour and of the relationship of this behaviour to the real situations which they
70 seek to represent, and the consequential modification of them so as to lead to reliable prediction and then to decision-taking, would not be possible were it not[4] for the assistance afforded to the investigator by the digital computer—and by the work of the technologists who have successfully transformed the scienti-
75 fic ideas on which it is founded[5] into stable, reliable and economic pieces of electrical equipment.

Where this new tool of investigation will ultimately take us is beyond my powers of prediction. But the subject is only a few years old and with the improvements in electronic techniques
80 which may confidently be expected, and the rapidly increasing knowledge and understanding of the brain possessed by the medical profession, it would perhaps be unwise to forecast undue restrictions on the nature of the ultimate achievement.

JACKSON, SIR WILLIS, *Penguin Technology Survey 1966*, Penguin Books Ltd.

[1] *parameter:* quantity or measurement which varies in different cases.
[2] *fantasy:* product of the imagination only.
[3] *as yet:* up to now, so far.
[4] *were it not:* except.
[5] *to be founded:* to be based.

Assignments

1 In the above passage, the author attempts to give a brief explanation of what a digital computer is and how it works. You may have noticed that this explanation is not altogether satisfactory—probably because it is too short. Could you give a better explanation, which could be substituted for the first two paragraphs?

2 Taking the example of an instrument or apparatus which is used in the discipline you yourself are studying, explain clearly for the layman what it is, how it works, and some of its main uses.

3 Explain (a) possible uses for a computer in your own discipline, (b) other uses in science and daily life not mentioned in the passage.

4 Find out, and explain in English, the connection between an understanding of how the human brain works and computer technology (ll. 80–82.)

EXPERIMENTATION IN THE SOCIAL SCIENCES

(The author enumerates some of the factors which make it difficult for the sociologist to conduct experiments in the same way as his colleagues in the physical sciences. One of the reasons for this is the lack of subject-matter which can be conveniently handled, and in this connection he points out that the common practice of using schoolchildren, university students or volunteers as subjects for experimentation—precisely because of their convenience—may lead to highly misleading results. This is because these classes of persons are atypical of the population as a whole, and the fact that the experiments, though often very accurately measured, are carried out in artificial conditions. He then goes on to describe the following experiments as examples of investigations carried out in more realistic conditions.)

EXPERIMENT I

The experimenter was himself a Socialist candidate for political office in a town. His purpose was to compare the effect of a 'logical' appeal for Socialism with that of an 'emotional' approach, so he prepared two contrasting leaflets.

5 The rational leaflet contained seven brief statements, to each of which the reader could signify agreement or disagreement. For example, one read: 'All banks and insurance companies should be run on a non-profit basis like the school.' At the end, the leaflet read:

10 Now go back and count the number of sentences with which you AGREED. Then count the number with which you DISAGREED. If the number of agreements is larger than the number of disagreements, you are at heart a Socialist—whether you know it or not. Why don't you try voting for the 15 thing you actually want? VOTE SOCIALIST!

The emotional appeal consisted of a letter addressed to 'Dear Father and Mother' and signed 'Your Sons and Daughters'. It was frankly sentimental, and ended with the words:

Our generation cannot enjoy the beauty and justice of the 20 New America if you block our highest desires. There was a time when you were young like us. We beg you in the name of those early memories and hopes to support the Socialist ticket in the coming elections! VOTE SOCIALIST!

The town had nineteen wards (electoral districts). The 25 emotional appeal was distributed to every family in three wards, the rational appeal to every family in four wards, while the remaining twelve wards served as controls, receiving neither. The distribution of incomes was much the same in the wards receiving the emotional appeal as in the wards receiving the rational 30 appeal.

The total vote in the city was one-sixth higher than in the pre-

ceding year, but the total Socialist vote—though still small—increased by nearly a third (31%). Increase was sharpest (50%) in the 'emotional' wards, but was also above average (35%) in
35 the 'rational' wards. In the remaining twelve wards which received neither leaflet the increase was only 24%.

It is not 'news' of course that an emotional appeal is more effective than a rational one. But it is a scientific event of significance when such a common assumption is put to the test of a
40 deliberately devised experiment having to do with realistic behaviour of ordinary adults in a complex social situation.

(The second experiment is interesting inasmuch as it apparently disproves a commonly-accepted assumption):

EXPERIMENT II

The experimenter (S. C. Dodd) set out to discover the relationships between a programme of rural hygiene and the hygienic practices of the families that were supposed to benefit from it. A mobile clinic was trying to instil hygienic behaviour into forty
5 families in the village of Jib Ramli in Syria. It is normally assumed that educational campaigns such as this lower morbidity and mortality, and increase comfort and happiness, but although millions are spent annually throughout the world on preventive medicine, little attempt has been made to test this
10 assumption.

Dodd's method was to select another village which matched Jib Ramli on nine factors: geographic, demographic, historical, economic, religious, domestic, recreational and in sanitary conditions. This second village was not exposed to the hygienic
15 campaign, and was therefore available as a control.

These two villages were assessed for 'hygienic performance' at the beginning of the experiment and again two years later. The method of assessment was to compile a list of 77 key questions. These questions were those retained from an initial
20 list of 270 questions, and together they showed some correlation with objective indices, such as mortality, morbidity and longevity.

It was found at the reassessment, at the end of the two years' experiment, that Jib Ramli's score had increased by 20%.
25 Unfortunately for the normal assumption, however, the hygiene score of the control village in the same period increased nearly as much—i.e. by over 18%. Either something had gone wrong with the experiment, or else the health programme was a waste of time and money.

MADGE, J., *The Tools of Social Science*, Longmans.

Assignments

1 Do you agree that schoolchildren, university students and volunteers may often be bad samples to use in sociological investigations? Draw distinctions, give reasons and quote examples.

2 One of the characteristics of the conventional methods of scientific experimentation is that of *replication*, i.e. the capacity for an experiment or series of experiments to be repeated under the same conditions. Is this possible in the social sciences? Is this factor an essential one, as far as sociological experiments are concerned? Give reasons in full for your opinions.

3 Imagine you are the investigator concerned in Experiment I. Give an account of it from a personal point of view and in your own words, as though you were explaining it to an international symposium of sociologists (use 'we' as the subject of the talk).

4 Imagine you are conducting a survey whose object (aim) is to find out the real attitudes of English-speaking foreigners in your country towards a certain political party or policy (which you can choose for yourself). Desig· and conduct an interview in English, using your classmates to represent the English-speaking 'population'.

5 Make an attempt to draw up some of the 77 questions used in Experiment II to determine 'hygiene performance'.

6 Imagine you are a participant in an international meeting at which Exp. II has just been described. In view of the fact that the findings (results) are contrary to common assumptions, formulate questions which you would ask the experimenter in order to satisfy yourself that the experiment had been carried out correctly.

PROBABILITY

The mathematics to which our youngsters are exposed at school is, with rare exceptions, based on the classical yes-or-no, right-or-wrong type of logic. It normally doesn't include one word about probability as a mode of reasoning or as a basis for com-
5 paring several alternative conclusions. Geometry, for instance, is strictly devoted to the 'if-then' type of reasoning and so to the notion (idea) that any statement is either correct or incorrect.

However, it has been remarked that life is an almost continuous experience of having to draw conclusions from in-
10 sufficient evidence, and this is what we have to do when we make the trivial decision as to whether or not to carry an umbrella when we leave home for work. This is what a great industry has to do when it decides whether or not to put $50,000,000 into a new plant abroad. In none of these cases—and indeed, in
15 practically no other case that you can suggest—can one proceed by saying, 'I *know* that A, B, C, etc. are completely and reliably true, and therefore the inevitable conclusion is . . .' For there is another mode of reasoning, which does not say: 'This statement is correct, and its opposite is completely false,' but which
20 says: 'There are various alternative possibilities. No one of these is *certainly* correct and true, and no one *certainly* incorrect and false. There are varying degrees of plausibility—of probability—for all these alternatives. I can help you understand how these plausibilities compare; I can also tell you how reliable my
25 advice is.'

This is the kind of logic which is developed in the theory of probability. This theory deals with not two truth values—correct or false—but with all the intermediate truth values: almost certainly true, very probably true, possibly true, unlikely,
30 very unlikely, etc. Being a precise quantitive theory, it does not use phrases such as those just given, but calculates for any question under study the numerical probability that it is true. If the probability has the value of 1, the answer is an unqualified 'yes' or certainty. If it is zero (0), the answer is an
35 unqualified 'no', i.e. it is false or impossible. If the probability is a half (0·5), then the chances are even that the question has an affirmative answer. If the probability is a tenth (0·1), then the chances are only 1 in 10 that the answer is 'yes'.

It is a remarkable fact that one's intuition is often not very
40 good at estimating answers to probability problems. For example, how many persons must there be in a room in order that the odds be favourable—that is, better than even—that there are at least two persons in the room with the same birthday (born on the same day of the month)? Remembering that there are 365
45 separate birthdays possible, some persons estimate that there would have to be 50, or even 100, persons in the room to make the odds better than even. The answer, in fact, is that the odds are better than even when there are 23 persons in the room; with 40 persons, the odds are better than eight to one that at
50 least two will have the same birthday. Let us consider one more

example: Everyone is interested in polls, which involve estimating the opinions of a large group (say all those who vote) by determining the opinions of a sample. In statistics the whole group in question is called the 'universe' or 'population'. Now
55 suppose you want to consult a large enough sample to reflect the whole population with at least 98% precision (accuracy) in 99 out of a hundred instances: how large does this very reliable sample have to be? If the population numbers 200 persons, then the sample must include 105 persons, or more than half the
60 whole population. But suppose the population consists of 10,000 persons or 100,000 persons? In the case of 10,000 persons, a sample, to have the stated reliability, would have to consist of 213 persons: the sample increases by only 108 when the population increases by 9,800. And if you add 90,000 more to the
65 population, so that it now numbers 100,000, you have to add only 4 to the sample! The less credible this seems to you, the more strongly I make the point that it is better to depend on the theory of probability rather than on intuition.[1]

Although the subject started out (began) in the seventeenth
70 century with games of chance such as dice and cards, it soon became clear that it had important applications to other fields of activity. In the eighteenth century Laplace laid the foundations for a theory of errors, and Gauss later developed this into a real working tool for all experimenters and observers. Any measure-
75 ment or set of measurements is necessarily inexact; and it is a matter of the highest importance to know how to take a lot of necessarily discordant data, combine them in the best possible way, and produce in addition some useful estimate of the dependability of the results. Other more modern fields of
80 application are: in life insurance; telephone traffic problems; information and communication theory; game theory, with applications to all forms of competition, including business, international politics and war; modern statistical theories, both for the efficient design of experiments and for the interpretation
85 of the results of experiments; decision theories, which aid us in making judgements; probability theories for the process by which we learn; and many more.

WEAVER, W., a talk reprinted in *Think*, April 1961.

[1] It should be remembered that the figures given in the example quoted in ll. 54–66 refer to completely representative samples, i.e. those that reflect the total population in all the aspects under consideration—a very difficult matter in the human sciences. Eds.

Assignments

1 Ask your mathematics teacher to show you the mathematical demonstration of the 'birthday problem' referred to in ll. 40–50. Then explain it to your classmates in English. What is the probability if there are 50 people in the room.?

2 Explain how you think probability theory could be used in any of the activities listed in ll. 80–87.

18

SUBJECT *Physics, Astronomy*
TYPE OF WRITING *Summarized explanation and hypothesizing*

QUASARS AND THE NEW UNIVERSE

Quasi-stars are a new phenomenon in the universe, and everybody is wondering what they are. Their light seems too bright to come from any known physical process. They broadcast powerful radio waves which may vary in strength. Some of them lie
5 near the limits of observable space and time, and promise to provide a crucial test of rival theories of the universe.

Quasi-stars were discovered two years ago (i.e. in 1963) as a result of an effort to overcome the shortcomings (defects) of radio telescopes. Compared to optical telescopes, these are
10 blunt instruments. They can spot (locate) a radio star (a source of radio waves in the sky) but can give only the most general clues as to its distance or nature. Progress depends on identifying radio stars with some kind of object emitting visible light—but radio astronomers can give their optical colleagues only
15 rather imprecise directions as to where to look.

It occurred to Cyril Hazard, a radio astronomer working at Jodrell Bank in England, that the moon could help. By waiting for it to eclipse a radio star, and timing the eclipse very precisely, a much more accurate position could be calculated. In 1962
20 Hazard and his colleagues used the technique to obtain a close fix on a powerful radio star catalogued as 3C-273. They passed the details to Maarten Schmidt, a Dutch astronomer working with the 200-inch telescope at Mount Palomar. Schmidt directed his instrument at the spot, and found a strange bright
25 star. He examined its light in a spectroscope and got a shock—the indications were that it was 1,500 million light-years away.

When astronomers look at the sky, they expect to see either a star or a galaxy. But Schmidt realized that 3C-273 could not be either: any ordinary star would be invisible at that range; yet
30 it was 200 times brighter, and much smaller, than a galaxy should be at such a distance. It was the first 'quasi-stellar radio source' or 'quasar' to be identified.

Within a year, 35 of these astonishing new objects had been found, and debate now centres on two questions: What are
35 quasars? and: What do they reveal about the universe? They promise finally to settle the conflict between the Big Bang and the Steady State theories, between an evolving or an unchanging universe.

The starting-point for both theories is the discovery in the
40 1930s that all the galaxies appear to be receding from us. The more distant galaxies are receding fastest, so we live in a universe which is expanding like an infinite balloon. The Big Bang theory holds that this expansion is the outcome (result) of an explosion long ago of some primeval 'atom' of inconceivably
45 dense matter. The Steady State theory holds that the universe has always expanded and always will, while new matter is continuously created to fill up the gaps.

To test the two theories, it is necessary to look far back in time. If the universe is evolving, it should have looked different
50 in the distant past—the galaxies must have been closer together.

If it is in a steady state, it must have looked just as it does now. When astronomers look out into the universe they see it not as it is, but as it was. We see the sun as it was 8 minutes ago, the nearest star as it was 4 years ago, the nearest galaxy as it was
55 2 million years ago. Quasar 3C-273 is 1,500 million years away, Quasar 3C-2 is 10,000 million years away and Quasar 3C-9 appears to be more distant still. They are thus revealing an incredibly ancient aspect of the universe.

Counts of the limited number of quasars so far identified
60 indicate that they become more abundant the more distant they are in space and time. In other words, the universe was different in the past, and is evolving: the Steady State theory must be dropped (discarded). Fred Hoyle, one of the original authors of the theory, all but (nearly) abandoned it this autumn: and he
65 cited as evidence against it not only the quasars, but another strange discovery of the past year.

Two Americans, Arno Penzias and Robert Wilson, were working last spring with an antenna used for satellite communications. Using ultra-sensitive equipment, they found they
70 were picking up (detecting, receiving) a constant soft radio whisper, which could not come from any known terrestrial or cosmic source. Similar observations have now been made at Princeton University. They put forward an extraordinary suggestion to account for it: their antenna, they said, was detecting
75 faint surviving energies from the original Big Bang, a background of ancient radio waves which now permeates the whole universe. It appears, Hoyle commented, that this mysterious whispering must have generated when the universe was denser than it is now—hence there has been evolution.

80 Meanwhile, there is the difficulty of how the quasars generate their enormous energies. One of the earliest ideas was that quasars are massive objects collapsing catastrophically inwards under the pull of their own gravity. This has not stood up to (survived) examination, but it has led to some still stranger
85 speculations. It appears that bodies undergoing gravitational collapse might be more or less invisible, since their huge masses, and the speed of the matter rushing inwards, would have the effect of bottling up their energies so that none could escape. It is thus possible that the universe contains numbers of huge
90 invisible bodies, detectable only by their gravitational fields. At the same time, space-borne astronomical instruments are beginning to reveal other unexpected aspects of the universe—huge cool stars which emit only infra-red rays, powerful sources of X-rays and ultra-violet radiation, and other phenomena normally
95 hidden by the earth's atmosphere.

It is becoming apparent that the universe is a very much more complex and surprising place than even the astronomers had suspected.

DAVY, J. *The Observer*, 19 December 1965.

Assignments

1 Give a simple explanation in English (suitable, for example,

for a child who did not understand the meaning of such terms as 'eclipse') of the problem of finding radio stars optically, and how Hazard solved it.

2 Explain clearly, so that a non-scientist could understand, the concepts of mass and energy, and hence the reason why bodies which undergo gravitational collapse might become invisible (ll. 85–88).

3 Find out, and describe to your classmates in simple English, the latest information in this field. The rest of the class should then ask questions on points arising from the talk.

Appendices and Basic Dictionary

Appendix A

COMMON PREFIXES AND SUFFIXES

PREFIX OR SUFFIX	MEANING OR FUNCTION	EXAMPLE
a- (an-)	not, not having	atypical (not typical); anhydrous (not containing water)
-able	forms adj from v	reliable (able to be relied on)
aero-	air	aerate (to force air through, e.g. a liquid)
-age	(1) forms n from v	storage (act of having stored)
	(2) forms abstract n with idea of an aggregate	tonnage (total number of tons)
-al	(1) forms adj from n	mathematical (belonging to mathematics)
	(2) forms n of action from v	trial (action of trying or testing)
-an (see **-ian**)		
-ant (-ent)	forms n and adj from v	resistant (capable of resisting); determinant (that which determines)
anti-	against	anti-toxin (substance acting against toxins)
-ate	(1) in the shape of, like	dentate (in the shape of a tooth)
	(2) possessing	nucleate (having a nucleus)
auto-	self, by itself	autogamic (self-fertilizing)
bi-	two	bi-metallic (consisting of 2 metals)
bio-	life	biology (science of life)
centi-	100 or $\frac{1}{100}$	centimetre ($\frac{1}{100}$ of a metre) Centrigrade (temperature scale with base of 100°)
co-	together, with	co-worker (person who works with someone else)
contra- / **counter-**	against, opposite	contra-rotating (rotating in 2 opposite directions); counteract (to act against, neutralize)
-cy	forms n from adj	accuracy (quality of being accurate)
de-	taken away from	dehydrated (with the water taken out)
deci-	a tenth	decimetre (tenth of a metre)
deka-	ten	dekametre (ten metres)
di-	two, twice	dioxide (compound containing 2 oxygen atoms)
dis-	not	disconnected (not connected)
-ent (see **-ant**)		
-er	forms n from v	transmitter (person or thing which transmits)

PREFIX OR SUFFIX	MEANING OR FUNCTION	EXAMPLE
hydro-	(1) water	hydrology (science of water in all its forms)
	(2) hydrogen	hydrocarbon (compound of H and C)
hyper-	over, excessive	hypertension (excessive blood pressure)
hypo-	below, less than usual	hypotension (subnormal blood pressure)
-ian	(1) forms n from sciences	statistician (person studying statistics)
	(2) forms n from countries	Canadian (person from Canada)
-ify	forms v from n or adj	intensify (to make intense)
in-	not	inaccurate
-ine	forms adj from n	saline (having the property of salt)
infra-	below, under	infra-red (below the wave-length of red)
inter-	between	interconnection (connection between)
intra-	inside	intravenous (inside the veins)
-ion (see **-tion**)		
-ish	a bit, resembling	reddish (a bit red)
iso-	equal	isostatic (equally balanced)
-ist	forms n from sciences	biologist (person who studies biology)
-ity	forms abstract n from adj	rigidity (quality of being rigid)
-ive	forms adj from v	selective (that which selects)
-ize	forms v from adj	standardize (to make standard)
kilo-	a thousand	kilogram(me) (1000 gram(me)s)
-less	forms adj from n	weightless (without weight)
-logy	study, science	geology (earth science)
-ly	forms adv from adj	slowly (in a slow manner)
macro-	large, on a large scale	macromolecule (large molecule)
mega(lo)-	very large, a million	megawatt (a million watts)
-ment	forms n from v	development (process of developing)
meta-	change	metamorphic (changed in form)
-meter	instrument which measures	thermometer (instrument which measures heat)
micro-	small, on a small scale	microclimate (climate in a small zone or area)
milli-	a thousandth	milligram(me) (1000th of a gm)

PREFIX OR SUFFIX	MEANING OR FUNCTION	EXAMPLE
mis-	badly, mistakenly	miscalculate (to calculate badly)
mono-	one, single	monochrome (of only one colour
multi-	many	multilateral (with many sides)
neo-	new	neolithic (belonging to the New Stone Age)
-ness	forms abstract n from adj	completeness (quality of being complete)
non-	not	non-conductor (substance which does not conduct electricity)
-oid	like, tending towards	anthropoid (like a man)
out-	(1) more than	outwear (wear or last longer than)
	(2) beyond, outside	outlying (beyond the main body)
over-	(1) more than, excessive	overproduction (too much production)
	(2) on top of, above	overlie (to lie on top of)
para-	similar to, irregular	paratyphoid (disease similar to typhus but of different origin)
pent(a)-	five	pentagon (5-sided figure)
-phono-	sound	phonology (science of speech sounds)
phot(o)	light	photometer (instrument for measuring the intensity of light)
poly-	many	polymorphous (having many shapes)
pre-	before, previously	pre-Cambrian (before the Cambrian (geological) Age)
proto-	first, original	prototype (first of a type or series)
quadri-	four	quadrivalent (having a valency of 4)
re-	(1) again, back	re-combine (to combine again after being separated)
	(2) together, mutually	react (to act on each other)
-scope	instrument for seeing	microscope (instrument for seeing small things)
self-	by itself	self-regulating (mechanism, etc., which regulates itself)
semi-	half, imperfect	semi-conductor (substance which does not conduct electricity very well)

PREFIX OR SUFFIX	MEANING OR FUNCTION	EXAMPLE
-sion (see **-tion**)		
-sis	process, state (in medicine a diseased state)	symbiosis (state of two different organisms living together)
sub-	under, below, less than	sub-atomic (below the size of atoms)
super- **(supra-)**	above, beyond, more than	supersonic (more than the speed of sound)
syn- (m-)	with, together	synthesis (process of putting together)
tetra-	four	tetrad (element having a valency of 4)
therm-	heat	thermometer (instrument for measuring heat)
-tion **(-sion)**	forms n from v	combination (result of combining)
tri-	three	triangle (figure with three angles)
-ty (see **-ity**)		
ultra-	beyond, more than usual	ultramicroscope (microscope showing smaller objects than the normal optical microscope)
un-	not	uneven (not even)
under-	(1) less than, insufficient	underpowered (having an insufficient power supply)
	(2) below, lower than	undersea (below the surface of the sea)
uni-	one	unicellular (having only one cell)

Appendix B

1 IRREGULAR VERBS

(The following list includes only the irregular verbs most frequently met with in scientific English)

INFINITIVE	PAST TENSE	PAST PARTIC.
arise	arose	arisen
be	was	been
bear	bore	born
become	became	become
bend	bent	bent
bind	bound	bound
break	broke	broken
breed	bred	bred
bring	brought	brought
build	built	built
choose	chose	chosen
come	came	come
cut	cut	cut
deal	dealt	dealt
do	did	done
draw	drew	drawn
drive	drove	driven
fall	fell	fallen
feed	fed	fed
find	found	found
fly	flew	flown
freeze	froze	frozen
get	got	got
give	gave	given
go	went	gone
grow	grew	grown
have	had	had
hold	held	held

INFINITIVE	PAST TENSE	PAST PARTIC.
keep	kept	kept
lead	led	led
leave	left	left
let	let	let
lie	lay	lain
light	lit	lit
lose	lost	lost
make	made	made
mean	meant	meant
meet	met	met
offset	offset	offset
overcome	overcame	overcome
put	put	put
ring	rang	rung
rise	rose	risen
run	ran	run
see	saw	seen
set	set	set
sink	sank	sunk
shake	shook	shaken
split	split	split
stand	stood	stood
take	took	taken
undergo	underwent	undergone
understand	understood	understood
wear	wore	worn
write	wrote	written

2 THE VERB 'TO BE'

PRESENT	PAST	FUTURE	FUTURE SUBSTITUTE[1]	CONDITIONAL
I am	I was	I shall (will) be	I am going to be	I would be
He, she, it is	He, etc. was	He, etc. will be	He is going to be	He would be
You are	You were	You will be	You are going to be	You would be
We are	We were	We shall (will) be	We are going to be	We would be
They are	They were	They will be	They are going to be	They would be

3 THE VERB 'TO HAVE'

PRESENT	PAST	FUTURE	FUTURE SUBSTITUTE[1]	CONDITIONAL
I have	I had	I shall (will) have	I am going to have	I would have
He, she, it has	He had	He will have	He is going to have	He would have
You have	You had	You will have	You are going to have	You would have
We have	We had	We shall (will) have	We are going to have	We would have
They have	They had	They will have	They are going to have	They would have

[1] This structure ('going to' plus the infinitive) is a substitute for the Simple Future which is more common in *spoken* scientific English than in the written form. It is sometimes called 'the Future of Intention', as it is held to express the intention to do something rather than mere futurity.

Appendix C

COMMON ABBREVIATIONS

(The following abbreviations include only those which are common to most or all of the various scientific disciplines. They do not include abbreviations used in particular sciences.)

A.C.	alternating current
Appx.	Appendix (to a publication)
approx.	approximately
av.	average
$\because$	because
Bul(l).	Bulletin
C	Centigrade (Celsius)
c.	about, approximately
ca.	about, approximately
cf.	compare
Ch.	Chapter
$^{\circ}$	degrees
D.C.	direct current
E	East
ed.	edited by
e.g.	for example
esp.	especially
$=$	equal, is equal to
$\simeq$	is approximately equal to
F	Farenheit
fig.	figure
ft.	foot (plural = feet)
′	feet; minutes
gm.	gram(me)
Hndbk.	handbook
hr.	hour
i.e.	that is to say
id.	the same
inf.	below
infra	below
in.	inches
″	inches; seconds
Jour.	Journal
km.	kilometres
l(ltr)	litre ('liter' U.S.A.)
$<$	less than (smaller than)
$\rightarrow$	leads to, is converted into
m.	metres
min.	minutes
ml.	miles
$>$	more than (larger than)
N	North
No.	number
#	number
op. cit.	the (literary) work already mentioned
p.	page
/	per (e.g. 8 km/sec. 8 kilometres per sec.)
p.c.	per cent
%	per cent
Pt.	Part (of a publication)
$\propto$	proportional to
pub.	published by
Publ.	Publication
Proc.	Proceedings
q.v.	which (you should) see
ref.	reference
S	South
sec.	second
Soc.	Society
sq.	square
sup.	above, before
supra	above, before
Suppl.	Supplement
T	temperature
$\therefore$	therefore
Trans.	Transactions
vide	see
Vol.	Volume
W	West
wt.	weight
yd.	yards

Appendix D

1 ENGLISH AND AMERICAN WEIGHTS AND MEASURES, AND APPROXIMATE METRIC EQUIVALENTS

WEIGHT		
1 ounce (oz.)		28 gm.
16 ozs.	1 pound (lb.)	450 gm.
112 lbs.	1 hundredweight (cwt.)	50 kgm.
20 cwt.	1 ton (tn.)	1000 kgm.[1]
LENGTH		
1 inch (in.)		2·5 cm.
12 ins.	1 foot (ft.)	30 cm.
3 ft.	1 yard (yd.)	91 cm.
6 ft.	1 fathom (f.)	1·83 m.[2]
1760 yds.	1 mile (ml.)	1·6 km.
VOLUME		
1 pint (pt.)		0·55 l.
2 pts.	1 quart (qt.)	1·1 l.
4 qts.	1 gallon (gal.)	4·5 l.[3]
AREA (AGRICULTURE)		
1 acre (ac.)		0·4 ha.
TEMPERATURE:		
32°F		0°C
212°F		100°C

[1] i.e. a 'long' ton; a 'short' or metric ton is 2000 lbs (910 kgm.).
[2] Used only in nautical or oceanographical work.
[3] A U.S. gallon is about 10% less than an English gallon.

2 SHAPES, ETC.

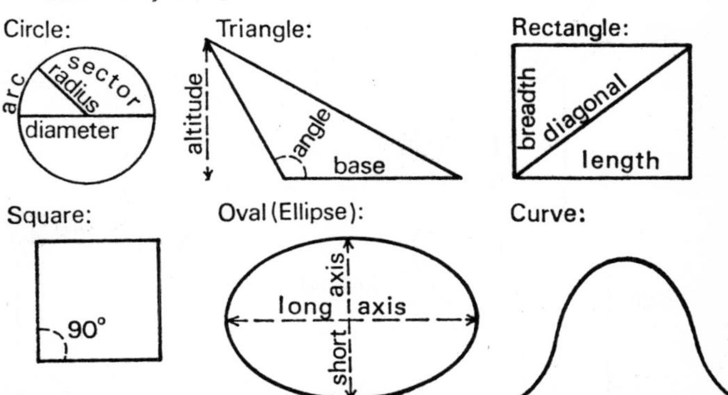

3 COLOURS

Using coloured pencils, colour the boxes below to correspond with their labels:

	White		Yellow
	Black		Orange
	Grey		Red
	Violet		Brown
	Light Blue		Green

NOTE: Approximate tones or shades of colours are indicated by prefixing the adjectives *light*, *medium* or *dark* as appropriate.

4 PRINCIPAL PARTS OF A HUMAN BODY AND OF A TREE

(These are frequently used in sciences other than biology, mainly in a metaphorical sense)

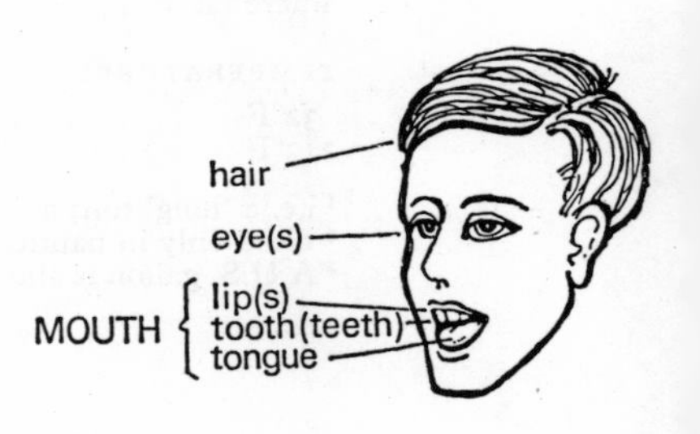

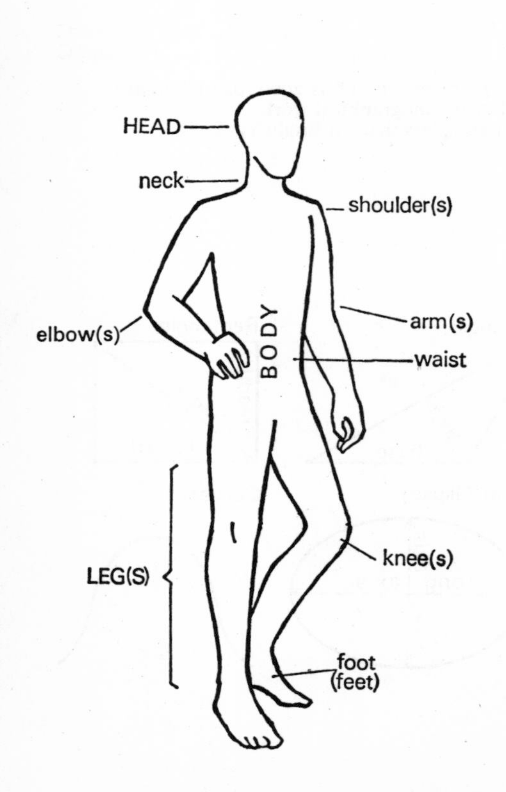

The eyes, nose, mouth, etc. are *features* of the *face*. The whole of the body is covered with *skin*.

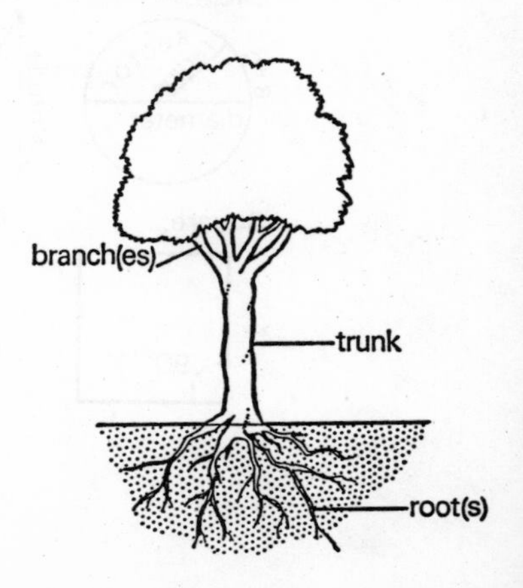

Dictionary of Basic Scientific English

The aim of the following dictionary is not to explain every word used in the course, but only those presented in the units which are common to all the main scientific disciplines, and therefore of basic importance for the understanding of the English of science as a whole.

PART I gives explanations of the non-structural items, using a limited vocabulary; it also lists, and gives the corresponding references to, all other items dealt with in Part II of the dictionary or in the Appendices.

PART II lists the main basic structural words and phrases. No explanations have been given for these items, since experience shows that such explanations, in general, fail to convey to the non-advanced student a clear idea of their meanings and the ways they are used. Instead, each item has been used illustratively within a context formed by the general basic scientific vocabulary of Part I.

In both Parts I and II, a space has been provided immediately to the left of each entry. This space may be utilized by the student for adding either a guide to pronunciation (in any convenient notation) or the vernacular equivalent, or both.

Abbreviations used in the Dictionary:

(*adj*)	adjective
(*n*)	noun
(*v*)	verb
(q.v.)	refer to the corresponding entry
(*)	irregular verb—refer to APPENDIX B
(agr.)	agriculture
(biol.)	biology
(chem.)	chemistry
(econ.)	economics
(eng.)	engineering
(geol.)	geology
(math.)	mathematics
(med.)	medicine
(phys.)	physics
(psych.)	psychology
(soc.)	sociology

NOTE:

Different meanings for the same functions of a word are separated by semicolons (;)—similar meanings are separated by commas (,).

Part I

A **a-** (see APPENDIX A).

ability (*n*) condition of being able, mental or physical power to do something, capacity.

-able (see APPENDIX A).

abnormal (*adj*) not normal (q.v.).

about (see PART II (General) under *Number and Quantity*).

above (see PART II (General) under *Position*).

above all (see PART II (Modifying Connectives) under *Emphasis*).

absent (*adj*) not present, lacking (q.v.).

absolute (*adj*) not relative (q.v.) or measured by comparison with other quantities; *absolute alcohol:* 100% pure alcohol; *absolute temperature* (*A°*): temperature at which molecular motion stops.

accelerate (*v*) to increase the speed of, to become or cause to become faster; (phys.) to change velocity (q.v.).

acceleration (*n*) increase of speed; (phys.) increase or decrease of speed, or change in the direction, of a body in motion.

in **accordance** with (see PART II (General) under *Others*).

according to (see PART II (General) under *Others*).

accordingly (see PART II (Modifying Connectives) under *Result*).

account (*n*) a description of a series of events; *on account of* (see PART II (Modifying Connectives) under *Result*).

account for (*v*) to give a satisfactory reason for; to make up a total of.

accumulate (*v*) to gather together; to become greater by addition, to increase in bulk or numbers.

accuracy (*n*) exactness, freedom from error.

accurate (*adj*) exact, free from mistakes or error.

acid (*n*) substance which turns litmus paper from blue to red and has a pH value of less than 7, i.e. a chemical compound containing hydrogen, in which the hydrogen can be partly replaced by a metal to form a salt.

across (see PART II (General) under *Movement*).

act (*n*) what is done.
(*v*) to have an effect (on); to perform the function of.

action (*n*) process of doing something or acting.

active (*adj*) not passive (q.v.); acting in an energetic way.

actually (see PART II (Modifying Connectives) under *Emphasis*).

adapt (*v*) to modify (q.v.), especially so as to be suitable for another purpose.

add (*v*) to cause an increase by joining one thing to another.

addition (*n*) process or result of adding.

in **addition** (see PART II (Modifying Connectives) under *Additional Statements*).

adequate (*adj*) sufficient, having the necessary qualities.

adjust (*v*) to alter or change so as to make something more efficient, or the right size, etc. for the purpose.

in **advance** (see PART II (General) under *Time*).

advantage (*n*) quality which makes something more useful or favourable.

aero- (see APPENDIX A).

affect (*v*) to act upon, so as to cause change or alteration.

after (see PART II (General) under *Position* and *Time*).

afterwards (see PART II (General) under *Time*).

against (see PART II (General) under *Position*).

age (*n*) length of time during which something has existed; (geol.) special division of geological time e.g. the Pleistocene Age.
(*v*) to become old.

-age (see APPENDIX A).

agent (*n*) that which acts so as to affect something.

aim (*n*) purpose (q.v.).

-al (see APPENDIX A).

all over (see PART II (General) under *Position*).

allot (*v*) to distribute (q.v.).

allow (*v*) to let, to permit; *to allow for:* to take into account, to include in the total calculation.

almost (see PART II (General) under *Degree*).

along (see PART II (General) under *Movement*).

already (see PART II (General) under *Time*).

also (see PART II (Modifying Connectives) under *Additional Statements*).

alter (*v*) to make or become different, to change or be changed.

alteration (*n*) state or process of being altered.

alternate (*adj.*) first one and then the other, arranged or coming one after the other by turns.
(*v*) to follow, or cause to follow one after the other by turns; *alternating current:* a flow of electricity which reverses its direction at regular intervals of time.

alternatively (see PART II (Modifying Connectives) under *Modifications*).

always (see PART II (General) under *Frequency*).

among (st) (see PART II (General) under *Position*).

amount (*n*) quantity, total sum of.

amount to (*v*) to add up to; to mean as much as.

an- (see APPENDIX A).

-an (see APPENDIX A).

analyse (-ze) (*v*) to separate something into its various parts in order to find out their values, functions, etc.

analysis (*n*) process or result of analysing.

-ance (see APPENDIX A).

and (see PART II (Modifying Connectives) under *Additional Statements*).

and so on (see PART II (General) under *Others*).

angle (*n*) the amount of slope of one line in relation to another, measured in degrees (see also APPENDIX D, under *Shapes, etc*); point of view.

-ant (see APPENDIX A).

anti- (see APPENDIX A).

any (see PART II (General) under *Number and Quantity*).

apart from (see PART II (Modifying Connectives) under *Additional Statements*).

apparatus (*n*) things, specially instruments, required for carrying out scientific operations and experiments.

apparently (see PART II (General) under *Others*).

appear (*v*) to come into view, to become visible; to seem, to give the impression (of, that).

appearance (*n*) action of appearing; visible characteristics of something.

apply (*v*) to put into use or action; to be relevant, to concern.

application (*n*) act or process of applying.

approach (*n*) a way or method of dealing with a problem; act of coming near.
(*v*) to come near; to begin to deal with a problem.

appropriate (*adj*) suitable for the purpose.

approximate (*v*) to be or come near to; to be nearly exact. (*adj*) not exactly, but almost; approaching correctness.

approximately (see PART II (General) under *Number and Quantity*).

approximation (*n*) a result not exactly correct but sufficient for some purposes.

arc (see APPENDIX D under *Shapes, etc.*).

area (*n*) amount of surface; field of action or study; region or zone (q.v.).

arise (**v*) to come into being or activity; to originate.

arm (see APPENDIX D under *Parts of the Human Body*).

around (see PART II (General) under *Position* and *Number and Quantity*).

arrange (*v*) to put in order.

arrangement (*n*) the act or result of arranging.

arrive (*v*) to get to a certain place; to reach.

as (see PART II (Modifying Connectives) under *Result*).

as far as (see PART II (General) under *Movement*).

as far as . . . is concerned (see PART II (General) under *Others*).

as follows (see PART II (General) under *Others*).

as regards (see PART II (General) under *Others*).

as soon as (see PART II (General) under *Time*).

as well as (see PART II (Modifying Connectives) under *Additional Statements*).

aspect (*n*) appearance (q.v.); point of view from which a subject can be analysed.

assemble (*v*) to put individual parts together to form a whole.

assembly(/)assemblage (*n*) a collection or association of inter-dependent things; *assembly line:* process of assembling industrial components into a finished product.

assess (*v*) to estimate (q.v.).

assist (*v*) to help.

assume (*v*) to accept as true without proof.

assumption (*n*) what is assumed.

at (see PART II (General) under *Time*).

-ate (see APPENDIX A).

atom (*n*) smallest part of an element (q.v.) which can take part in a chemical reaction.

atomic (*adj*) belonging to or connected with atoms.

attach (*v*) to fasten, to join to; *to attach importance to:* to consider important.

attain (*v*) to reach, to arrive at.

attempt (*n*) an effort to do something; (*v*) to make an effort to do something.

attract (*v*) to pull towards by means of a force (e.g. magnetism).

attraction (*n*) state of process of attracting.

auto- (see APPENDIX A).

automatic (*adj*) working by itself without external control.

available (*adj*) which can be used or obtained.

average (*n*) result obtained by adding a group of unequal quantities together and dividing its total by the number of quantities. (*v*) to obtain an average; to maintain or reach an average rate of; (*adj*) property belonging to an average; usual, ordinary.

avoid (*v*) to try not to do or make.

away (see PART II (General) under *Position*).

B

back (*n*) part of a thing away from or farthest from the observer; opposite of front (q.v.) (see also PART II (General) under *Position* and *Movement*).
(*adj*) belonging to the back.

backward(s) (see PART II (General) under *Movement*).

balance (*n*) laboratory apparatus for measuring weight; the state which exists when two opposing forces are equal.
(*v*) to keep in a state of balance.

base(is) (*n*) part which serves as foundation or main support; the main part or component in a mixture; a substance which turns litmus paper from red to blue and has a pH value of more than 7 and which reacts with an acid to form a salt and water.
(*v*) to build or place upon; to take as a starting-point.

basic (*adj*) belonging to a base; absolutely necessary.

beam (*n*) long horizontal piece of wood, steel, etc. used in construction engineering; a stream of light, magnetic or electrical emissions.

bear in mind (**v*) to consider, to remember.

bearing (have a **bearing** on) (*n*) (to have) a connection or relation with, an influence on.

because (see PART II (Modifying Connectives) under *Result*).

become (**v*) to come to be; to develop into.

before (see PART II (General) under *Position* and *Time*).

beforehand (see PART II (General) under *Time*).

in the **beginning** (see PART II (Modifying Connectives) under *Order of Events*).

behave (*v*) to act, to function.

behaviour (*n*) state or process of behaving or reacting in the given circumstances.

below (see PART II (General) under *Position*).

bend (*n*) a curve (q.v.), a curved angle.
(**v*) to form or cause to form a curve or curved angle; to deflect.

beneath (see PART II (General) under *Position*).

beside (see PART II (General) under *Position*).

besides (see PART II (Modifying Connectives) under *Additional Statements*).

between (see PART II (General) under *Position*).

beyond (see PART II (General) under *Movement*).

bi- (see APPENDIX A).

bind (**v*) to fasten together.

bio- (see APPENDIX A).

bit (*n*) a small piece, a fraction of a whole; *a bit:* slightly.

black (see APPENDIX D under *Colours*).

block (*v*) to obstruct, to stop the normal flow of.

blockage (*n*) an obstruction, a stoppage in the normal flow.

blue (see APPENDIX D under *Colours*).

body (*n*) *human body* (see APPENDIX D under *Parts of the Human Body*); any object having mass; institution or organization.

boil (*v*) to heat a liquid until its vapour pressure is the same as that of its immediate surroundings.

bond (*n*) anything which binds (q.v.), especially molecules (chem.).

bottom (*n*) lowest part, base.
(*adj*) belonging to the lowest part or base (see also PART II (General) under *Position*).

to be **bound** up with (*v*) to be inevitably related to or connected with.

branch (*n*) (see APPENDIX D under *Parts of a Tree*). Extension or subdivision of a large whole.
(*v*) to divide or become divided into branches.

break (*n*) an interruption, a gap.
(**v*) to interrupt or cause to be interrupted; to separate or be separated into bits, usually violently; *to break up:* to break into several or many separate bits

break down (**v*) to fail, to stop functioning; to separate into component parts in an orderly and controlled process.

breakdown (*n*) a failure, stoppage.

breed (*n*) specially-developed variety of living organism.
(**v*) to reproduce or cause to reproduce in a controlled and selective way.

bridge (*n*) engineering structure used to take a road or railway over an obstacle; means of communication between two otherwise separate things (e.g. fields of study, ideas, etc.).
(*v*) to establish a means of communication between two separate things.

bright (*adj*) full of light.

bring (**v*) to cause to come from a farther place to one nearer.

brown (see APPENDIX D under *Colours*).

build up (**v*) to increase or cause to increase by regular additions.

bulk (*n*) size; main quantity of.

but (see PART II (Modifying Connectives) under *Modification*).

by (see PART II (General) under *Time*).

by-product (*n*) something additional obtained as a result of the development or production of something else.

C

calculate (*v*) to obtain a result by making use of mathematical procedures.

calculation (*n*) process or result of calculating.

capable (*adj*) having the capacity or ability to do something.

capacity (*n*) power or ability to do something; space to hold or to contain, usually measured in liquid units (e.g. litres, gallons, etc.).

carry (*v*) to support the weight of something and at the same time to move it from one place to another.

carry out (*v*) to bring to a successful conclusion, to perform completely.

case (*n*) an example or instance; all the circumstances of a scientific investigation, especially in medicine, psychology and sociology.

cause (*n*) anything which produces an action in something else.
(*v*) to produce an action or effect (q.v.).

cell (*n*) the smallest unit of living matter capable of existing independently; a device for producing electricity by chemical means.

Celsius (see **centigrade** below).

cent (i)- (see APPENDIX A).

centigrade (Celsius) (*n* & *adj*) temperature scale in which the melting-point of ice is taken as 0° and the boiling-point of water as 100°.

central (*adj*) middle; most important.

centre (*n*) the middle part; important thing.

certain (*adj*) having no doubt about, sure; some but not much or many; specific but not described in detail.

certainly (see PART II (Modifying Connectives) under *Emphasis*).

chain (*n*) series of things, ideas or events connected together.

chance (*n*) an opportunity, a favourable time for something to act or happen; random or accidental occurrence.
(*adj*) random, accidental.

change (*n*) state or process of becoming, or causing to become, different.
(*v*) to become or cause to become different; to replace or substitute, to cause to replace or substitute.

channel (*n*) a path for the transmission of electrical signals; any means through which ideas, etc. are communicated.

characteristics (*n*) the special qualities or properties which make up the identity of anything.

charge (*n*) quantity of electricity present.
(*v*) to cause something to have a charge.

chart (*n*) (see **graph** below), map used for hydrographical purposes.

check (*n*) anything which stops or reduces the speed of a process; process of controlling in order to ensure accuracy.
(*v*) to stop or reduce the speed of a process; to control in order to ensure accuracy.

choice (*n*) the act of selecting between two or more alternatives or possibilities.

choose (**v*) to select between two or more alternatives or possibilities.

circle (see APPENDIX D under *Shapes, etc.*).

circulate (*v*) to move or flow continuously so as to return to the starting-point; to move freely.

circulation (*n*) process of circulating.

class (*n*) a group having the same or similar characteristics.

clear (*adj*) obvious, free from doubt; easy to see through, transparent; easy to understand; free from obstruction.
(*v*) to make clear.

clear-cut (*adj*) definite, well-defined.

clearly (see PART II (Modifying Connectives) under *Emphasis*).

close (*v*) to shut or cause to shut; to come to an end.
(*adj*) near at hand, not distant.

co- (see APPENDIX A).

cold (*n*) low temperature; (med.) inflammation of nose caused by virus.
(*adj*) low in temperature, lacking heat.

collapse (*v*) to fall to pieces; to fail, to break down; (med.) to lose strength suddenly.
(*n*) state or process of collapsing.

colour (see APPENDIX D under *Colours*).

combine (*v*) to mix or cause to mix in a controlled way.

combination (*n*) the act or result of combining.

common (*adj*) usual, frequent; belonging to or used by more than one.

communicate (*v*) to transmit or exchange information.

communication (*n*) act or process of communicating.

compare (*v*) to examine or look for similarities and differences in two or more things.

comparative (*adj*) which involves comparisons.

comparatively (see PART II (General) under *Degree*).

comparison (*n*) the act or process of comparing.

complete (*v*) to finish, to bring to a conclusion; to add so as to make a whole.
(*adj*) finished, concluded; whole.

completely (see PART II (General) under *Degree*).

completion (*n*) act or process of completing.

complex (*n*) a group or assemblage (q.v.).
(*adj*) consisting of many closely-connected parts; involved, complicated.

component (*n*) item which helps to make up a whole.

compose (*v*) to make up, to form.

compound (*n*) a mixture, especially (chem.) of chemical elements (q.v.) bound together in definite proportions and having a specific arrangement of atoms.
(*adj*) made up of two or more parts.

concentrate (*v*) to increase the amount of an element or compound in a solution (q.v.); to group together, to bring together at one point.

concentration (*n*) act or process of concentrating.

concern (*v*) to relate to, to affect (q.v.); to be of interest to; *as far as . . . is concerned* (see PART II (General) under *Others*).

conclude (*v*) to end, to finish; to form an opinion or judgement after considering the evidence or facts.

conclusion (*n*) end; opinion, or judgement formed after considering the evidence.

condense (*v*) to concentrate (q.v.), to reduce to a smaller volume; to change from vapour to liquid.

condition (*n*) state of being or existing; necessary factor without which something else cannot happen.
(*v*) to modify, to act or cause to act in such a way as to alter the behaviour of; to limit, to restrict.

conduct (*n*) behaviour (q.v.).
(*v*) to perform, to carry out; to allow (heat, electricity, etc.) to pass from one place to another.

conform (*v*) to act or behave in accordance with; to be in agreement with.

conformity (*n*) act or process of conforming; (geol.) undisturbed relationships of geological strata.

connect (*v*) to join or be joined; to relate or be related with.

consequence (*n*) result (q.v.); *of consequence:* important.

consequently (see PART II (Modifying Connectives) under *Result*).

conserve (*v*) to maintain unchanged, to keep from loss or wastage.

consider (*v*) to think carefully.

consist (*v*) to be made up or composed of.

consistent with (see PART II (General) under *Others*).

constant (*adj*) which does not alter or change, which remains the same.
(*n*) a quantity or factor which does not change.

constitute (*v*) to compose (q.v.), to form by the addition of its parts.

construct (*v*) to build; to put together parts to make a whole.

construction (*n*) thing constructed; process of constructing.

consume (*v*) to use up (q.v.).

contain (*v*) to have space for; to enclose; to include.

contents (*n*) that which is contained.

contra- (see APPENDIX A).

contract (*v*) to become or cause to become shorter or smaller.

on the **contrary** (see PART II (General) under *Others*).

control (*v*) to keep within limits; to exert authority or power over something;
(*n*) factor or group of factors in an experiment which provide the basis for comparison with other factors being measured (variables); device for exerting power over something.

conversely (see PART II (General) under *Others*).

conversion (*n*) act or process of converting.

convert (*v*) to change from one form into another.

cool (*adj*) somewhat cold.
(*v*) to become or cause to become cool; to remove heat from.

co-ordinate (*v*) to relate or bring together various factors so as to increase their joint efficiency.

correct (*adj*) not wrong; appropriate.
(*v*) to set right, to make correct.

correlate (*v*) to bring one thing into the correct relationship with others; to show the relationships existing between various things.

correspond (*v*) to be equivalent to.

count (*v*) to say the numerals in order; to be equivalent to; to be important.
(*n*) the act of saying the numerals in order; total number; *on various counts:* for various reasons.

counter- (see APPENDIX A).

counterbalance (*v*) to exert a force or influence of equal strength to.
(*n*) that which counterbalances.

of **course** (see PART II (Modifying Connectives) under *Emphasis*).

cover (*v*) to put or spread something over something else; to include.
(*n*) that which is put over something else.

crop (*n*) plants grown for food; total quantities of such plants grown in a single season.

cross (*v*) to go or move from one side to another; to mix the genetic units of plants or animals in order to produce better varieties.
(*n*) the result of crossing plants or animals; *cross-section* (see **section** below).
(*adj*) moving or pointing in two or more different directions—e.g. cross-bedding (geol.), cross-current (metereology, oceanography); referring from one thing to another (e.g. cross-reference).

cross- (see APPENDIX A).

cumulative (*adj*) increasing by successive additions.

current (*n*) a stream of liquid, gaseous or electrical particles flowing in the same direction.
(*adj*) of the moment, of the present.

curve (*n*) (see APPENDIX D under *Shapes, etc.*).
(*v*) to be in the shape of a curve.

cut (**v*) to divide into two or more parts, to separate; to stop (e.g. machinery); to make a narrow opening by means of a sharp instrument.
(*n*) result of cutting; a reduction.

cut down (**v*) to reduce, to make less.

-cy (see APPENDIX A).

cycle (*n*) a process of transformation which comes back to the same point from which it began and is then repeated.

D

damage (*n*) loss caused by partial or total destruction of elements essential to the functioning of anything.
(*v*) to cause damage to.

damp (*adj*) not completely dry, slightly wet.

dark (*adj*) having little or no light; a deep shade (of colours), i.e. reflecting light of a lower frequency.

data (*n*) relevant facts in any given problem.

de- (see APPENDIX A).

deal with (**v*) to do what is necessary; to explain; to solve.

a great **deal** (see PART II (General) under *Number and Quantity*).

decade (*n*) period of ten years.

deci- (see APPENDIX A).

decimal (*n*) fraction of a number expressed in a non-fractional notation based on tenths, hundredths, etc.
(*adj*) belonging to a decimal; *decimal system:* a system of measurement based on multiples and fractions of ten.

decrease (*v*) to become or cause to become less in number or quantity.
(*n*) process of decreasing; amount by which something is decreased.

deduce (*v*) to reach conclusions after considering the relevant facts.

deduct (*v*) to take away, to subtract.

deduction (*n*) what is deduced; what is deducted.

deep (*adj*) extending far downwards, inwards or across; marked, profound.

defect (*n*) fault, factor that causes failure.

deficient (*adj*) insufficient, not having enough.

deficiency (*n*) quality of being deficient.

define (*v*) to describe in accurate terms.

definition (*n*) result or process of defining; (optics) sharpness or clearness of image obtained.

degree (*n*) relative amount or intensity; unit of various scales used to measure temperatures, angles and geographical positions.

deka- (see APPENDIX A).

demonstrate (*v*) to show, usually by giving proof.

delay (*v*) to be or cause to be late or slow.
(*n*) result of delaying.

dense (*adj*) having the component parts very close together.

density (*n*) quality of being dense; (phys.) the mass of a substance per unit of volume: $\frac{M}{V}$

depend on (*v*) to need the special support of.

depth (*n*) distance downwards, inwards or across.

deposit (*v*) to lay down, to spread over.
(*n*) layer of material deposited.

design (*v*) to plan in a careful and accurate manner.
(*n*) a careful and accurate plan.

destroy (*v*) to break to pieces; to put an end to.

destruction (*n*) result or process of destroying.

detect (*v*) to succeed in finding something whose presence is difficult to discover.

determine (*v*) to set limits to; to define (q.v.); to be the cause of, to be the most important factor in.

develop (*v*) to grow, advance or increase, or cause to grow, advance or increase; to advance through successive stages to a higher or more complete state, to evolve; (photography) to treat a film with chemicals so that a usually negative picture is obtained.

development (*n*) process of developing.

deviate (*v*) to change direction owing to external circumstances; to differ from an accepted standard or norm.

device (*n*) piece of apparatus.

di- (see APPENDIX A).

diagonal (*n*) (see APPENDIX D under *Shapes, etc.*).
(*adj*) (see APPENDIX D under *Shapes, etc.*).

diagram (*n*) simplified drawing which shows how something functions.

differ (*v*) to be unlike.

difference (*n*) quality of being unlike; specific point at which things differ.

different (*adj*) unlike, not the same; various.

differentiate (*v*) to make or point at differences between one thing and another.

difficult (*adj*) not easy; requiring a lot of effort or skill to do.

difficulty (*n*) that which is difficult; quality of being difficult.

dimension (*n*) measurement; (usually *dimensions*) size.

diminish (*v*) to be or make less or smaller.

dip (*v*) to be or move downwards at an angle below the horizontal. (*n*) the angle at which something is or moves downwards below the horizontal.

direct (*v*) to cause to move in a certain direction; to control. (*adj*) straight, uninterrupted; *direct current:* (electricity) one which flows continuously in one direction only.

direction (*n*) course taken by a moving object at any given moment; act of directing.

dis- (see APPENDIX A).

discover (*v*) to find out; to get knowledge of (often for the first time).

dissolve (*v*) to cause a substance to divide into very small particles in a liquid, thus causing a solution (q.v.); to break down, to disintegrate.

distant (*adj*) not near, far away.

distance (*n*) amount of space or time between two objects or events.

distribute (*v*) to spread out or disperse over time or space; to give out proportionately.

distribution (*n*) state or process of distributing or being distributed

disturb (*v*) to alter the normal condition of.

disturbance (*n*) something that disturbs; act or process of disturbing something.

diverge (*v*) to move in different directions from a common point; to vary from a common norm or standard.

divergence (*n*) the amount by which something diverges; act of diverging.

divide (*v*) to separate or to become separated into two parts; (math.) to find out how many times a number is contained in another.

division (*n*) act or process of dividing.

double (*v*) to multiply by two; to make or be made twice as big, etc. (*adj*) twice the amount; compound; consisting of two identical items.

down (see PART II (General) under *Movement*).

downwards (see PART II (General) under *Movement*).

drift (*v*) to move or be moved passively, at random. (*n*) passive, random movement; very slow movement of large masses of land or water.

drive (*v*) to direct or control the operation or movement of; to give movement to by means of a mechanism. (*n*) a controlling or directing device or mechanism; a mechanism which produces movement.

drop (*v*) to let fall; to decrease or grow less. (*n*) a small quantity of liquid; a decrease or fall.

due to (see PART II (Modifying Connectives) under *Result*).

during (see PART II (General) under *Time*).

E

easily (see PART II (General) under *Degree*).

easy (*adj*) not difficult (q.v.).

edge (*n*) line marking the outer limit of a (flat) surface; sharp, cutting part of a knife or similar instrument.

effect (*v*) to cause.
(*n*) a result; a particular phenomenon in science, usually associated with the name of its discoverer, e.g. Stark effect (spectroscopy).

effective (*adj*) able to cause the required result.

efficient (*adj*) effective with the minimum amount of effort.

efficiency (*n*) being efficient; (mechanics) ratio between the input of energy and the eventual output, usually expressed as a percentage.

effort (*n*) amount of force used.

either . . . or (see PART II (General) under *Others*).

elapse (*v*) to pass (of time).

electricity (*n*) movement of electrons (at present it cannot be adequately defined).

electronic (*adj*) having to do with the technological applications of electrons and their behaviour.

element (*n*) one of a little over 100 substances each composed of a single type of atom having a specific number and arrangement of electrons, from which all other substances are built up; one of the component factors making up a complex thing or subject; *elements:* main outline of a (complex) subject.

eliminate (*v*) to take out or remove inessential or undesirable elements.

emerge (*v*) to appear (gradually).

emit (*v*) to give out or send out.

empty (*adj*) having nothing inside, containing nothing.

enable (*v*) to cause to be able.

end (*n*) farthest or last part or point, termination.
(*v*) to come to or cause to come to or reach an end.
(*adj*) final, at the end of a series.

energy (*n*) (mechanics) capacity for doing work; (phys.) that which is transferred or transformed in a physical or chemical reaction; power.

enlarge (*v*) to make larger.

enough (see PART II (General) under *Degree*).

-ence (see APPENDIX A).

-ent (see APPENDIX A).

entirely (see PART II (General) under *Degree*).

equal (*adj*) having the same value and/or characteristics as.
(*v*) to be equal to.

equation (*n*) a mathematical statement indicating equality between two expressions involving known and unknown quantities; symbolical representation of a chemical reaction.

equipment (*n*) collection of apparatus or other objects necessary for a given purpose.

equivalent (*adj*) having the same value or function as; capable of producing the same results as.

-er (see APPENDIX A).

especially (see PART II (General) under *Degree*).

essential (*adj*) absolutely necessary.

establish (*v*) to set up, to instal(l); to put on a firm basis.

estimate (*v*) to make an approximate calculation.
(*n*) that which is estimated.

etcetera (etc.) (see PART II (General) under *Others*).

evaluate (*v*) to estimate the value or effects of.

event (*n*) anything that happens.

eventual (*adj*) coming at last as a result.

eventually (see PART II (General) under *Time* and (Modifying Connectives) under *Order of Events*).

evidence (*n*) facts necessary for proving something.

evident (*adj*) clear, obvious.

exact (*adj*) accurate (q.v.).

exactly (see PART II (General) under *Degree*).

examine (*v*) to observe or think about with care and accuracy.

example (*n*) fact, thing or person chosen as representative of a group to illustrate a general rule; *for example* (see PART II (General) under *Others*).

exceed (*v*) to go beyond a certain limit; to be more than what is expected or required.

exceedingly (see PART II (General) under *Degree*).

except (*v*) not to consider, to consider as an exception; (see PART II (Modifying Connectives) under *Modification*).

exception (*n*) that which does not conform to a general rule.

excess (*n*) result of exceeding.

exchange (*v*) to give or receive one thing in place of another.
(*n*) act or process of exchanging.

exclude (*v*) not to consider or contain as part of.

exclusively (see PART II (General) under *Restriction*).

exert (*v*) to apply (e.g. pressure or force).

exhaust (*v*) to use up all the strength or resources of; to empty (e.g. a gas from a chamber).

exist (*v*) to be; to be found.

expand (*v*) to grow from a smaller size to a larger one.

expansion (*n*) result or process of expanding.

expect (*v*) to consider as likely to happen.

expense (*n*) money, time or effort spent; *at the expense of:* causing a proportional amount of loss to.

experiment (*n*) test or procedure scientifically carried out in order to gain new knowledge.
(*v*) to carry out a series of experiments.

explain (v) to make plain and clear by words or mathematical symbols; to account for, to give a satisfactory reason for.

explanation (*n*) result or process of explaining.

exploit (*v*) to utilize the resources of.

extend (*v*) to enlarge, to expand, to cover an area of.

extension (*n*) result or process of extending.

extensively (see PART II (General) under *Degree*).

extent (*n*) degree, amount; area; *to a great (large, small, etc.) extent* (see PART II (General) under *Degree*).

external (*adj*) being or belonging to the outside.

extreme (*adj*) of the highest degree or furthest limits.
(*n*) *extremes:* very wide variation (e.g. of temperature, etc.).

extremely (see PART II (General) under *Degree*).

eye (*n*) (see APPENDIX D under *Parts of the Human Body*).

F

face (*n*) (see APPENDIX D under *Parts of the Human Body*).
(*v*) to be in a position looking towards the face of; to meet directly, to encounter as an obstacle.

fact (*n*) event or phenomenon known to exist or have existed.

in **fact** (see PART II (Modifying Connectives) under *Emphasis*).

factor (*n*) fact or series of facts or conditions having an influence on; (math.) one of the numbers which when multiplied together produce a given number (e.g. 5 and 2 are the factors of 10); (eng.) *safety factor:* amount of extra strength beyond what is calculated as the minimum necessary in the given conditions.

Fahrenheit (*n & adj*) temperature scale in which the melting-point of ice is taken as 32° and the boiling-point of water as 212°.

fail (*v*) not to be successful; to break down, to be unable to continue; to be inadequate or incomplete.

failure (*n*) the result of failing.

fair (*adj*) moderately good, not very bad.

fairly (see PART II (General) under *Degree*).

fall (*v*) to move rapidly or steeply from a higher to a lower position; to decrease, to grow less, to drop.
(*n*) process of falling.

false (*adj*) not true; misleading.

as **far** as (see PART II (General) under *Movement*).

so **far** (see PART II (General) under *Time*).

fast (*adj*) quick, rapid; able to move from one place to another in a short time; (of colours) fixed.

feature (*n*) (see APPENDIX D under *Parts of the Human Body*); a characteristic (q.v.).

a **few** (see PART II (General) under *Number and Quantity*).

female (*adj*) belonging to the sex which bears young.
(*n*) a member of the sex which bears young.

field (*n*) (phys.) area over which a force (e.g. magnetism) operates or exerts an influence; area of activity or study; area of land producing crops or minerals.
(*adj*) *field-work, field studies, field conditions:* those taking place or occurring in non-laboratory environments.

fill (*v*) to make or become full, to occupy the whole space.

finally (see PART II (Modifying Connectives) under *Order of Events*).

find (**v*) to obtain as a result of systematic search or investigation.

finding(s) (*n*) results or conclusions obtained from investigation.

at **first** (see PART II (Modifying Connectives) under *Order of Events*).

fit (*v*) to be the correct size or shape for; to correspond exactly with.
(*adj*) suitable, appropriate.
(*n*) way in which one thing fits another, adjustment.

fix (*v*) to be or cause to be unchanging or stable; to attach one thing to another so that it cannot move; to decide, to determine (e.g. prices).

flat (*adj*) having a level or horizontal surface.

flatten (*v*) to make flat.

flight (*n*) act or process of flying (q.v.).

flow (*v*) to move easily and continuously, like a liquid.
(*n*) process of flowing; that which flows.

fluctuate (*v*) to increase and decrease in an irregular way, to vary.

fluctuation (*n*) result or process of fluctuating.

fluid (*n*) a liquid; that which flows.
(*adj*) moving freely, flowing; liquid.

fly (**v*) to move or be moved through the air; to leave, to abandon.
(*n*) small flying insect.

follow (*v*) to come after or next in time or space; *it follows that* (see PART II (Modifying Connectives) under *Result*); *as follows* (see PART II (General) under *Others*).

foot (*n*) (see APPENDIX D under *Parts of the Human Body*); (see APPENDIX D under *Weights and Measures*).

for (see PART II (Modifying Connectives) under *Result*).

force (*n*) (phys.) external agency capable of altering the state of rest or motion of a body; power or energy exerted.
(*v*) to compel, to make something or somebody do something.

forecast (*v*) to estimate in advance what is likely to happen.
(*n*) act or process of forecasting.

form (*n*) general appearance, usually external; a variety or type.
(*v*) to make up, to compose.

the **former** (*n*) the first-named of two.

formula (*n*) a condensed statement in words or (more frequently) symbols expressing a relationship between two phenomena.

forwards (see PART II (General) under *Movement*).

fraction (*n*) (math.) a part less than an integer, expressed in terms of a numerator and a denominator, e.g. $\frac{2}{3}$ (two-thirds); a bit, less than the whole.

free (*adj*) separate, not combined with or attached to anything else; not subject to external control.
(*v*) to make free.

freeze (**v*) to convert from liquid to solid by lowering the temperature; to immobilize, to make unable to move.

frequent (*adj*) occurring at short intervals.

frequency (*n*) (phys.) number of vibrations per unit time; number of times an event occurs in a given unit of time or space.

frequently (see PART II (General) under *Frequency*).

front (*n*) the part facing the observer; the part facing the direction of motion; (meteorology) line marking the limits between two different masses of air.
(*adj*) belonging to the front (see also PART II (General) under *Position*).

full (*adj*) containing as much as the capacity allows; complete.

fully (see PART II (General) under *Degree*).

function (*n*) special purpose; *to be a function of:* to depend upon and vary with.
(*v*) to perform, to act.

fundamental (*adj*) basic, essential, very important; forming the bottom part of.

furthermore (see PART II (Modifying Connectives) under *Additional Statements*).

future (*n*) the time coming after the present.
(*adj*) coming after the present time.

G

gap (*n*) empty space or interval of time between two objects or events.

gas (*n*) least dense state of matter within any given range of temperature and pressure.

general (*adj*) concerning or affecting all or almost all; not specific; widespread in time and space.

generally (see PART II (General) under *Frequency*).

given (see PART II (Modifying Connectives) under *Result*).

give off (**v*) to emit.

give rise to (**v*) to cause, to originate.

go on (**v*) to continue.

government (*n*) process of directing or controlling a country; body of men who direct or control a country.

gradual (*adj*) developing by slow degrees or stages.

graduate (*v*) to mark with degrees for measuring; to obtain a university degree.

graph (*n*) a line or lines drawn to represent the relationship of one quantity to another.

gray (grey) (see APPENDIX D under *Colours*).
green (see APPENDIX D under *Colours*).
gross (*adj*) whole, total, without deductions (q.v.).
grounds (*n*) reasons.
on the **grounds** of (see PART II (General) under *Others*).
group (*n*) a number or set of things having common characteristics or behaviour.
(*v*) to arrange in a group; to classify.
grow (**v*) to increase in size, quantity or complexity; (agr.) to cultivate (e.g. crops).
growth (*n*) result or process of growing.

H **half** (*n*) when a whole is divided into two exactly equal parts, each part is known as a half (pl. hal*ves*).
halve (*v*) to divide into halves.
hand (*n*) (see APPENDIX D under *Parts of the Human Body*); *on the other hand* (see PART II (Modifying Connectives) under *Modification*).
handle (*v*) to deal with; to use, to operate.
hard (*adj*) not easily cut or broken; unyielding to pressure; difficult.
harden (*v*) to make or become hard.
hardly (see PART II (General) under *Degree*).
hardly ever (see PART II (General) under *Frequency*).
harm (*n & v*) damage (q.v.).
head (see APPENDIX D under *Parts of the Human Body*).
heat (*n*) degree of intensity of molecular movement, accompanied by a rise in temperature.
(*v*) to cause or to have a rise in temperature.
heavy (*adj*) having more weight (q.v.) than a norm or what is usual.
height (*n*) the vertical component of size; measurement from top to bottom.
hence (see PART II (Modifying Connectives) under *Result*).
here (see PART II (General) under *Position*).
high (*adj*) having more height (q.v.) than a norm or what is usual; having more intensity than a norm.
highly (see PART II (General) under *Degree*).
hinder (*v*) to put obstacles in the way of.
hitherto (see PART II (General) under *Time*).
hole (*n*) empty space or concavity on or through a solid body.
hollow (*adj*) having an empty space inside.
(*n*) broad, shallow concavity on the surface of something.
horizontal (*adj*) parallel to the earth's surface (see also PART II (General) under *Position*).
hot (*adj*) having more heat than the norm or what is usual.
how (see PART II (General) under *Others*).
however (see PART II (Modifying Connectives) under *Modification*).
huge (*adj*) extremely large.
hydro- (see APPENDIX A).
hyper- (see APPENDIX A).
hypo- (see APPENDIX A).
hypothesis (*n*) a provisional explanation of a phenomenon; an unproved theory.

I **-ian** (see APPENDIX A).
identify (*v*) to make sure or recognize the exact nature or scientific

category of; (psych.) *to identify with:* to behave in a similar way to someone with whom one has strong emotional bonds.

identity (*n*) specific nature; that which makes one thing different from another.

if (see PART II (Modifying Connectives) under *Condition*).

-ify (see APPENDIX A).

imagine (*v*) to think, to form a mental picture of.

imagination (*n*) ability or capacity to imagine; ability to think creatively and inventively.

imaginative (*adj*) quality of using imagination.

imply (*v*) to have as a further consequence.

importance (*n*) quality of being important.

important (*adj*) having great value, influence or effect.

improve (*v*) to make or become better.

improvement (*n*) act or process of improving; something which improves something else; something which results from the process of improving.

in (see PART II (General) under *Position* and *Time*).

in advance (see PART II (General) under *Time*).

in- (see APPENDIX A).

inasmuch as (see PART II (Modifying Connectives) under *Result*).

include (*v*) to contain or to consider as a part of.

increase (*v*) to grow or cause to grow in number, quantity and/or size.
(*n*) growth in number, quantity and/or size.

indeed (see PART II (Modifying Connectives) under *Emphasis*).

indicate (*v*) to provide evidence from which further conclusions may be drawn; to consider or be considered necessary; to show.

individual (*adj*) belonging to or characteristic of a single member of a group; single, by itself.

-ine (see APPENDIX A).

infer (*v*) to deduce (q.v.).

inference (*n*) result of inferring.

inferior (*adj*) low in quantity or value.

influence (*n*) power or capacity to affect.
(*v*) to exert an influence on.

information (*n*) relevant knowledge, data; (cybernetics) signals and their transmission for the purpose of controlling processes.

infra- (see APPENDIX A).

in front of (see PART II (General) under *Position*).

initially (see PART II (Modifying Connectives) under *Order of Events*).

inner (see PART II (General) under *Position*).

-ion (see APPENDIX A).

input (*n*) amount of energy, material or information received.

inside (see PART II (General) under *Position*).

inside out (see PART II (General) under *Position*).

in so far as (see PART II (Modifying Connectives) under *Result*).

instance (*n*) example, case, *for instance* (see PART II (General) under *Others*).

instrument (*n*) apparatus or device for doing a special job in science or technology with a high degree of accuracy.

insulate (*v*) to prevent loss of energy (e.g. heat, electricity, etc.) from a substance by surrounding it with a non-conductor.

intake (*n*) input (q.v.).

integrate (*v*) to combine parts into a whole.
intend (*v*) to have the purpose of.
intense (*adj*) having a high degree or concentration of.
inter- (see APPENDIX A).
internal (*adj*) of or belonging to the inside.
interval (*n*) length of time or space between two events or things.
into (see PART II (General) under *Movement*).
intra- (see APPENDIX A).
introduce (*v*) to bring into use or operation for the first time.
investigate (*v*) to examine scientifically and systematically.
investigation (*n*) act or process of investigating.
involve (*v*) to have or bring as a consequence; to become or cause to become connected with.
inwards (see PART II (General) under *Movement*).
irreversible (*adj*) which cannot be reversed (q.v.).
-ish (see APPENDIX A).
iso- (see APPENDIX A).
isolate (*v*) to separate from others or from its surroundings.
-ist (see APPENDIX A).
-ity (see APPENDIX A).
-ive (see APPENDIX A).
-ize (see APPENDIX A).

J

job (*n*) specific piece of work.
join (*v*) to be or cause to be together or connected with others.
(*n*) place or surface where two or more things join.
joint (*n*) a join; (biol.) a structure which enables two bones to fit together.
(*adj*) done by two or more persons or organizations.
jump (*n*) sudden rise, acceleration or increase; a sudden movement from one place to another.
(*v*) to make a jump.
just (see PART II (General) under *Time*).

K

keep in mind (*v) (see **bear in mind**).
key (*n*) something which provides a solution or a method of approach to a problem.
(*adj*) very important.
kilo- (see APPENDIX A).
kind (*n*) class or group having the same characteristics which differ from those of other groups.
knowledge (*n*) cumulative and systematic data or information.

L

laboratory (*n*) room or series of rooms specially equipped for conducting scientific experiments.
lack (*v*) not to have, to be without.
(*n*) state of lacking.
lag (*n*) excessive delay between cause and effect.
lag (behind) (*v*) to be too slow or inadequate.
large (*adj*) big in size or amount.
largely (see PART II (General) under *Degree*).
at **last** (see PART II (Modifying Connectives) under *Order of Events*).
lastly (see PART II (Modifying Connectives) under *Order of Events*).
later on (see PART II (Modifying Connectives) under *Order of Events*).
the **latter** (*n*) the last-named of two.

law (*n*) statement of an apparently constant relationship between two phenomena under strictly defined conditions.

lay (*v*) to place in a horizontal position.

layer (*n*) a relatively thin sheet (q.v.) of any material spread over an area.

lead to (**v*) to have as a result.

least (*adj*) smallest in size, quantity or degree.
(*n*) the smallest.

at **least** (see PART II (General) under *Degree*).

left (-hand) (*adj*) on or of the side of the human body where the heart is located.

length (*n*) measurement from end to end (of space and time).

less (than) (*adj*) smaller in size, quantity or degree (than) (see also PART II (General) under *Number and Quantity*).

-less (see APPENDIX A).

let (**v*) to allow, to permit; (math.) to assume, to consider as.

level (*n*) horizontal surface; specified height or depth; specified standard; instrument for measuring the horizontal and angles from the horizontal.
(*adj*) horizontal.
(*v*) to make level or horizontal; (eng. and geol.) to establish the comparative levels of features of the surface of the land.

liable (*adj*) subject to the possibility.

light (*n*) visible radiation.
(**v*) to emit or cause to emit light.
(*adj*) not heavy (of weight (q.v.)); not dark (of colour).

likely (*adj*) probable.

limit (*n*) line or point marking the maximum extension or value of.
(*v*) to establish limits for, to restrict (q.v.)).

line (*n*) a long, thin mark made by pen, etc.; a series of objects one behind the other or one next to the other; (math.) form having only one dimension, e.g. length.
(*v*) to arrange objects in a line; to cover the inside of.

lip (*n*) (see APPENDIX D under *Parts of the Human Body*).

link (*n*) a connection; part of a chain (q.v.).
(*v*) to connect together.

a **little** (see PART II (General) under *Degree*).

liquid (*n*) state of matter which is neither solid nor gas.

load (*n*) amount of weight or force supported by a structure; electric power carried by a circuit; amount of power which is required from a machine at any given time.
(*v*) to cause to carry a load; (statistics) to give additional value to.

locate (*v*) to be or be put in a specific place.

-logy (see APPENDIX A).

long (*adj*) having length (q.v.); extending more in space or time than expected.

long-term (-range) (*adj*) operating or having an effect over a long extension of time or space.

loose (*adj*) not tight or fixed, able to move freely.

lose (**v*) to be unable to keep or retain.

loss (*n*) act, process, or result of losing.

a **lot** (see PART II (General) under *Number and Quantity*).

low (*adj*) having less height (q.v.) than the norm; small in quantity or intensity.

lower (*adj*) having the least height of two; being underneath of two.
(*v*) to cause to decrease or be less; to cause to move from a higher to a lower position.

-ly (see APPENDIX A).

M **machine** (*n*) device or apparatus composed of several or many parts working together in order to produce power and/or movement.

machinery (*n*) machines considered as a unit; institutions or organizations involved in a given process.

macro- (see APPENDIX A).

magnitude (*n*) degree of size, importance or intensity; (astronomy) relative brightness of stars, etc.

main (*adj*) most important.

mainly (see PART II (General) under *Degree*).

maintain (*v*) to keep, to hold; to keep in working condition (e.g. machinery); to hold a view supported by evidence.

maintenance (*n*) process of keeping in working condition.

major (*adj*) large, having great importance.

make up (**v*) to compose, to consist of.

male (*n & adj*) opposite of female (q.v.).

many (see PART II (General) under *Number and Quantity*).

mark (*n*) sign used for indicating or identifying.
(*v*) to make a mark.

marked (*adj*) obvious, easily seen, large.

mass (*n*) amount of matter in a body usually measured in terms of weight (q.v.); the majority; *mass media* or *mass-communication:* newspapers, the radio, television and the cinema.

material (*n*) that of which a thing is made; apparatus; *raw materials:* agricultural or mineral products from which manufactured goods are made.
(*adj*) essential, important, relevant.

matter (*n*) arrangement of atoms of which any physical object is made; subject.
(*v*) to be of importance.

maximum (*n*) the greatest possible amount, degree or intensity.

maybe (see PART II (General) under *Degree* and (Modifying Connectives) under *Doubt or Reservation*).

mean (*n*) average obtained by dividing the sum of a number of quantities by the number of quantities.
(*adj*) belonging to a mean.
(**v*) to express a certain idea; to imply (q.v.); to have as a purpose.

means (*n*) a method; *by means of* (see PART II (General) under *Others*).

meaning (*n*) that which is meant (q.v.), significance.

in the **meantime** (see PART II (General) under *Time*).

measure (*n*) (usually plural, *measures*): action(s) taken to achieve a specific purpose; that which is used to estimate the importance, etc. of something.
(*v*) to calculate exactly the dimensions or intensity of something.

measurement (*n*) act or result of measuring.

mechanical (*adj*) belonging to or acting like a machine.

mechanism (*n*) a series of interconnected processes which describe how something functions or has evolved.

medium (*n*) substance or environment surrounding an object or through which it acts.
(*adj*) neither big nor small, not extreme.

mega- (see APPENDIX A).

melt (*v*) to change from solid to liquid due to an increasing temperature.

(*n*) a mixture resulting when two or more solids melt or are melted together.

-ment (see APPENDIX A).

merely (see PART II (General) under *Restriction*).

meta- (see APPENDIX A).

(-)meter (*n*) an instrument for measuring (see also APPENDIX A).

metre (meter, USA**)** (see APPENDIX D under *Weights and Measures*).

method (*n*) systematic way of doing something.

micro- (see APPENDIX A).

middle (*adj*) point equally distant from the ends or limits of.

milli- (see APPENDIX A).

minimum (*n & adj*) the smallest possible amount, degree or intensity.

minor (*adj*) small; having little importance.

minus (*adj*) not having, without; (math. and phys.) having a negative (−) value or electrical charge.

minute (*n*) 60 seconds (of time or angle).
(*adj*) very small, tiny.

mis- (see APPENDIX A)

mistake (*n*) a deviation from accuracy or truth.
(*v*) to identify wrongly.

mix (*v*) to combine into a compound; to put different things, etc. together.

mixture (*n*) result of mixing.

model (*n*) a hypothesis, series of integrated concepts about how something behaves; (eng.) small-scale copy of a larger object.

moderate (*adj*) not extreme (q.v.), of medium size, etc.

moderately (see PART II (General) under *Degree*).

modify (*v*) to make changes in, especially improvements.

modification (*n*) result or process of modifying.

moisture (*n*) liquid present in a solid or gas, wetness.

molecule (*n*) smallest unit into which a chemical compound can be divided without losing its identity.

mono- (see APPENDIX A).

more (than) (*adj*) large in size, quantity and degree (than) (see also PART II (General) under *Time* and *Number and Quantity*).

moreover (see PART II (Modifying Connectives) under *Additional Statements*).

at (the) **most** (see PART II (General) under *Degree*).

motion (*n*) movement, process of changing position.

mount (*v*) to increase; to prepare and put in a fixed position for study (e.g. a laboratory specimen).

mouth (*n*) (see APPENDIX D under *Parts of the Human Body*).

move (*v*) to change or to cause to change position.

movement (*n*) process of moving.

much (see PART II (General) under *Number and Quantity*).

multi- (see APPENDIX A).

multiple (*n*) (math.) a number into which a smaller number can be divided an exact number of times.
(*adj*) having a number of different parts or forms.

multiply (*v*) to add a number to itself a given number of times; to increase rapidly in number.

multiplication (*n*) process of multiplying.

N **name** (*n*) word(s) by which anything is known.
(*v*) to give a name to.

narrow (*adj*) of little width in comparison with length; having less width than normal or expected; limited.
(*v*) to make narrow.

nature (*n*) the essential characteristics of anything; the universe and its phenomena.

natural (*adj*) belonging to nature; not made in the laboratory or factory.

naturally (see PART II (Modifying Connectives) under *Emphasis*).

near (see PART II (General) under *Position*).

nearly (see PART II (General) under *Degree*).

neck (*n*) (see APPENDIX D under *Parts of the Human Body*).

negative (*adj*) (math.) belonging to the class of numbers inferior to 0, and characterized by the minus sign (−); (phys.) belonging to the type of electricity characterized by the minus sign (−); (psych.) not constructive, passive in attitude.
(*n*) (photography) a developed film in which natural light and dark are reversed.

neighbouring (*adj*) near or next to.

neither . . . nor (see PART II (General) under *Others*).

neo- (see APPENDIX A).

-ness (see APPENDIX A).

net (*n*) network (q.v.).
(*adj*) remaining when all the necessary subtractions have been made.

network (*n*) a system of interconnected things or relationships.

neutral (*adj*) belonging to neither positive nor negative; not influenced either in favour or against.

never (see PART II (General) under *Frequency*).

nevertheless (see PART II (Modifying Connectives) under *Modification*).

next (see PART II (Modifying Connectives) under *Order of Events*).

non- (see APPENDIX A).

none (see PART II (General) under *Number and Quantity*).

norm (*n*) the ordinary or most frequent value or state; standard.

normal (*adj*) usual, belonging to the norm.

not only . . . but also (see PART II (General) under *Others*).

note (*v*) to observe, to fix the attention on.

notwithstanding (see PART II (Modifying Connectives) under *Modification*).

now (see PART II (General) under *Time*).

nowadays (see PART II (General) under *Time*).

nucleus (*n*) central or most important part round which other parts are grouped; (phys.) central part of an atom containing a positive electrical charge and making up most of the mass of the atom; (biol.) central part of a cell (q.v.) containing the genetic components.

nuclear (*adj*) belonging to a nucleus.

number (*n*) convention (usually expressed in units, tens and multiples and fractions of tens) by which things are counted.

O **object** (*n*) a thing.
(*v*) to give as a reason against.

objection (*n*) a reason given against.

objective (*n*) an aim (q.v.).
(*adj*) acting uninfluenced by external considerations; considering all the relevant evidence.

oblique (*adj*) at an angle between horizontal and vertical (see also PART II (General) under *Position*).

observation (*n*) act or process of observing.

observe (*v*) to look at, and note the behaviour of, systematically.

obstacle (*n*) something which prevents or makes difficult the attainment of an objective.

obtain (*v*) to get; to exist, to be present.

obviously (see PART II (Modifying Connectives) under *Emphasis*).

occasionally (see PART II (General) under *Frequency*).

occur (*v*) to happen, to be or be found.

occurrence (*n*) that which occurs, an event.

offset (*v*) to counterbalance the effects of.

often (see PART II (General) under *Frequency*).

-oid (see APPENDIX A).

on (see PART II (General) under *Position*).

once (see PART II (General) under *Time* and *Frequency*).

on top of (see PART II (General) under *Position*).

only (see PART II (General) under *Number and Quantity* and *Restriction* and also under **not only** above).

open (*adj*) opposite of closed or shut; allowing things to go in or out freely.

operate (*v*) to have an effect on; to function or cause to function; to perform an operation on.

operation (*n*) result or process of operating; (med.) medical procedure which involves the cutting of a diseased part of a living body; (math.) procedure involved in adding, subtracting, etc.

opposite (*adj*) facing (see also PART II (General) under *Position*); reverse, totally different.

optimum (*adj & n*) the best, the most suitable.

orange (see APPENDIX D under *Colours*).

order (*n*) systematic way in which things or events are arranged.

in **order to** (see PART II (General) under *Others*).

to be **of the order of** having the approximate magnitude of; more specifically, to the nearest power of ten or additional significant figure.

ordinary (*adj*) usual, normal, average.

organic (*adj*) concerned with or having the characteristics of an organism.

organism (*n*) individual capable of growth and reproduction; a whole made up of parts working together.

organize (*v*) to group or arrange into a system.

origin (*n*) the beginning, the starting-point.

original (*adj*) having to do with the origin of; new, not done before.

otherwise (see PART II (Modifying Connectives) under *Modification*).

out (see PART II (General) under *Position*).

out- (see APPENDIX A).

outline (*n*) line showing shape or limits of; statement of most important or most relevant facts.
(*v*) to make or give an outline.

output (*n*) total produced per unit time; opposite of **input** (q.v.).

at the **outset** (see PART II (Modifying Connectives) under *Order of Events*).

outside (see PART II (General) under *Position*).

outstanding (*adj*) very good or important.

outwards (see PART II (General) under *Movement*).

over (see PART II (General) under *Position, Time* and *Number and Quantity*).
over- (see APPENDIX A).
overall (*adj*) including all the main points but excluding minor details.
overcome (**v*) to be stronger than.
overhead (see PART II (General) under *Position*).
owing to (see PART II (Modifying Connectives) under *Result*).
own (*adj*) belonging to oneself, not belonging to other things or people.
oxygen (*n*) element O, the most frequently occurring in the earth's atmosphere and crust.

P

para- (see APPENDIX A).
parallel (*adj*) exactly similar or closely comparable to, having the same form or stages as, equivalent to; (math.) (of lines) keeping at the same distance from each other over their whole length.
part (*n*) a component; *to play a part:* to carry out a function.
partially (see PART II (General) under *Degree*).
particularly (see PART II (General) under *Degree*).
partly (see PART II (General) under *Degree*).
passive (*adj*) acted upon, but not itself acting.
pattern (*n*) interconnected arrangement of things or events.
peak (*n & adj*) maximum (q.v.).
pent(a)- (see APPENDIX A).
per (*prep.*) in or for each.
per cent (*n*) in or for every hundred units.
percentage (*n*) rate or number per cent; proportion, number or quaniity.
perfectly (see PART II (General) under *Degree*).
perform (*v*) to do, to carry out, to function.
performance (*n*) act or process of performing.
perhaps (see PART II (General) under *Degree* and (Modifying Connectives) under *Doubt and Hypothesis*).
permanent (*adj*) lasting or intended to last for an indefinite period.
perpendicular (*adj*) forming a right angle (90°) with a line or surface.
phase (*n*) stage of development; state of a body or quantity which passes repeatedly through a series of changes, or which varies periodically.
phenomenon (*n*) any thing or event in nature which can be a subject for study.
-phono- (see APPENDIX A).
photo- (see APPENDIX A).
piece (*n*) a part or fraction of a whole, a bit; an item, e.g. a piece of machinery, a piece of information.
place (*n*) particular part of a space occupied by something; position in any scale of value.
(*v*) to be or put into a place.
plain (*adj*) clear, easy to see or understand, obvious.
(*n*) area of flat land.
plan (*n*) a calculation of what must be done to achieve a certain purpose; a drawing or diagram (e.g. of a machine or building) made to scale.
(*v*) to make a plan.
plane (*n*) a perfectly level surface.
(*adj*) having a uniformly flat and level surface.

plant (*n*) living organism having leaves, flowers, etc. (usually excluding trees); machinery used in industrial processes or for producing power.

plot (*v*) to mark the position of something on a graph (q.v.).
(*n*) result of plotting; piece of land (often for building or agricultural experiments).

plus (*adj*) (math.) having a positive (+) value or electrical charge.

point (*n*) definite item or idea in an explanation or description.

point out (*v*) to show, to bring to the attention of.

pole (*n*) each of the ends of a magnet, including the earth (North and South Magnetic Poles); each of the ends of the axis of the Earth (the North and South Geographic Poles); each of the terminals of an electrical battery.

poly- (see APPENDIX A).

pool (*n*) a common store (of resources, etc.) shared by many.
(*v*) to form a pool.

poor (*adj*) low in quality; small in quantity.

population (*n*) all the things (especially people) in a given area; (statistics) any collection of individuals, things or measurements having some common characteristics.

positive (*adj*) opposite of negative (q.v.).

possibly (see PART II (General) under *Degree* and (Modifying Connectives) under *Doubt or Hypothesis*).

potential (*adj*) existing in undeveloped form, capable of being used or developed.
(*n*) that which is potential; (phys.) the work done in bringing a unit of mass, electricity, etc. from infinity to a given point in the field of force.

potentially (see PART II (General) under *Others*).

power (*n*) energy (q.v.); (math.) number of times a quantity is multiplied by itself; strength, force; capacity, ability; electricity, as in *power station*, *power consumption*, etc.

practical (*adj*) having industrial or technological application (often opposed to **theoretical,** q.v.).

pre- (see APPENDIX A).

precisely (see PART II (General) under *Degree*).

prepare (*v*) to get ready.

predict (*v*) to forecast (q.v.).

present (*adj*) being in the place expected, existing.
(*n*) the time now, at the moment of speaking or acting, or reading these words; *at present* (see PART II (General) under *Time*).
(*v*) to show; to give evidence of.

pressure (*n*) force exerted at a given point, measured in units of weight per unit area.

presumably (see PART II (General) under *Others*).

prevent (*v*) to stop from reaching a given objective.

primarily (see PART II (General) under *Degree*).

primary (*adj*) having first place or importance.

principal (*adj*) main, most important.

principle (*n*) basic law (q.v.).

problem (*n*) a matter difficult to make clear or solve.

procedure (*n*) a method or series of methods.

process (*n*) series of interconnected actions or methods taken as a whole, used to achieve a certain objective.
(*v*) to subject to a process; to arrange systematically and subject to analysis.

produce (*v*) to cause, to give rise to; to yield (q.v.); to make or manufacture.
(*n*) products (q.v.) especially agricultural.

product (*n*) that which is produced; (math.) result of multiplying; *by-product:* product of secondary importance obtained from the manufacture of something else; *end-product:* final product obtained as the result of a series of processes.

production (*n*) process of producing, amount produced.

profoundly (see PART II (General) under *Degree*).

program(me) (*n*) statement of plans (q.v.).
(*v*) to draw up a series of instructions to enable a computer to deal with a given problem.

progress (*n*) movement forwards; rate and extent of improvement.
(*v*) to make progress.

project (*n*) plans (q.v.); investigation (q.v.).
(*v*) (stats. & econ.) to plot the tendencies of a series of values into the future on the assumption that their known relationships will remain unchanged; (math.) to show a three (or more) dimensional object on a flat surface.

proof (*n*) evidence (q.v.).

property (*n*) a characteristic (q.v.).

proportion (*n*) percentage, fraction (q.v.); relationship in size., number, degree, etc. between one thing and another, ratio (q.v.).

proto- (see APPENDIX A).

prove (*v*) to produce evidence of the truth or validity of a statement.

provide (*v*) to give what is needed; *provided that* (see PART II (Modifying Connectives) under *Condition*).

provisional (*adj*) not permanent; intended to be changed or replaced later.

provisionally (see PART II (General) under *Others*).

pull (*v*) to cause to come towards.

pulse (*n*) (med.) sudden increase and decrease in the flow of blood in the arteries caused by the action of the heart; rate of this measured in number of fluctuations per minute; a regular increase and decrease in the magnitude or quantity (e.g. of electric current, radio waves, etc.) whose value is usually constant.

pure (*adj*) not mixed with any other substance.

purpose (*n*) something one tries to attain, intention.

push (*v*) to cause to go away from.

put forward (**v*) to present for examination (e.g. a theory).

Q

quadri- (see APPENDIX A).

quality (*n*) essential characteristic or property; degree of relative value.

quantity (*n*) amount (q.v.).

query (*v*) to feel and express a lack of certainty about.
(*n*) a question (q.v.); an uncertainty, a doubt; a problem.

question (*n*) something asked, which requires an answer; a problem, a query.
(*v*) to inquire into or about, to ask questions; to query.

quick (*adj*) fast (q.v.).

quite (see PART II (General) under *Degree*).

quote (*v*) to repeat something said or written by another person; (econ.) to state the current price for.

R

raise (*v*) to cause to move from a lower to a higher position; (agr.) to grow or produce crops or animals.

random (*adj*) made or done without aim or method, irregular; acted on by accidental forces.

range (n) maximum distance or area over which the effects can be felt; group or series of similar objects, usually arranged in order of size or intensity; distance between two extremes (e.g. temperature, etc.); a chain (q.v.) of mountains or hills.
(*v*) to arrange in a special order; to vary between definite limits.
(*adj*) *long* (*short*)-*range:* see under **long** and **short.**

rapid (*adj*) fast (q.v.).

rare (*adj*) not frequently-occurring.

rarely (see PART II (General) under *Frequency*).

rate (*n*) proportion (q.v.); speed; (econ.) percentage of interest charged by banks on loans.
(*v*) to give an approximate value to.

rather (see PART II (General) under *Degree*).

rather than (see PART II (General) under *Others*).

ratio (*n*) the relation between two or more quantities determined by the number of times the smallest one(s) is or are contained in the largest.

re- (see APPENDIX A).

reach (*v*) to succeed in arriving at, to extend as far as.

react (*v*) to act again; to act in the opposite direction to; to act in answer to a stimulus (q.v.).

reaction (*n*) result or process of reacting.

readily (see PART II (General) under *Degree*).

really (see PART II (Modifying Connectives) under *Emphasis*).

recent (*adj*) of, or happening in, the near past.

recently (see PART II (General) under *Time*).

recognize (*v*) to identify as known or met with previously; to be willing to accept, to see the truth of.

record (*n*) an accurate account usually in permanent form of one or a series of events or measurements.
(*v*) to make a record.

reduce (*v*) to make less, to diminish; (chem.) to remove oxygen from or add hydrogen to a substance; (math.) to simplify, to express in terms of larger units.

reflect (*v*) to give or send back (e.g. light, heat) directly or at an angle; to reproduce accurately.

refer (*v*) to direct the attention to, to point out.

reference (*n*) act or result of referring.

regardless of (see PART II (Modifying Connectives) under *Modification*).

as **regards** (see PART II (General) under *Others*).

with **regard to** (see PART II (General) under *Others*).

region (*n*) large area without definite boundaries.

register (*v*) to indicate (q.v.); to record (q.v.).

regular (*adj*) occurring at fixed or equally-spaced intervals.

relate (*v*) to show the connection between two or more phenomena.

relation(ship) (*n*) act or result of relating.

relative (*adj*) comparative (q.v.).

relatively (see PART II (General) under *Degree*).

release (*v*) to set free, to allow to go out; to loosen.

relevant (*adj*) having to do with, specifically connected with.

reliable (*adj*) which can be relied on.

reliability (*n*) state or degree of being reliable.

rely (on) (*v*) to have confidence in; to depend (up)on.

remove (*v*) to take away.

repeat (*v*) to do or say again.

represent (*v*) to be a sign of or a symbol for; to mean, to indicate.

research (*n*) investigation aimed at increasing knowledge about some particular subject and/or its technological application.

resist (*v*) to hinder or prevent the action of; not to be affected by.

resistance (*n*) act or power of resisting; (phys.) degree (measured in ohms) to which a material hinders the flow of an electric current.

resources (*n*) supplies; items available for use.

restrict (*v*) to limit, to prevent from reaching the maximum.

result (*n*) that which is produced by a cause or follows as a consequence or effect; answer to a (mathematical) problem.
(*v*) to have as a result.

retain (*v*) to keep back, to continue to have or to hold.

return (*v*) to come, go or send back, or to cause to do so.
(*n*) act or process of returning; (econ.) amount of money gained.

reveal (*v*) to show, to make known.

reverse (*v*) to cause to move, flow, etc. in the opposite direction; to turn upside down.

revolve (*v*) to move in a circle around a centre or axis.

revolution (*n*) act or process of revolving; (geol.) a period of major mountain building; (soc.) a period of great political and social change, usually occurring through the use of force.

right (*adj*) correct, true, valid; opposite of left (q.v.).

right-hand (*adj*) on or of the right (q.v.) side.

ring (*n*) a circular piece of metal or assemblage of objects in the shape of a circle (e.g. atoms in certain molecules); sound produced by a bell.
(*v*) to surround by a ring; to produce or cause to produce a ring.

rise (**v*) to move from a lower to a higher level or position; to increase.
(*n*) act or process of rising.

role (*n*) a function (q.v.).

root (*n*) (see APPENDIX D under *Parts of a Tree*); (math.) a fraction of a number such as when multiplied by itself it produces the whole number.

rough (*adj*) approximate (q.v.); having a broken, irregular surface.

roughly (see PART II (General) under *Degree*).

run (**v*) (of a machine, etc.) to function; (of an organization, etc.) to be responsible for, to be in charge of; (of experiments) to perform; (geol.) to extend in a certain direction.
(*n*) a series; *in the long* (*short*) *run:* after a long (short) time has elapsed.

S

safe (*adj*) free from danger.

sample (*n*) one, or more generally a group of items or quantity of substance taken as representative of a whole.
(*v*) to take samples.

save (*v*) to prevent or avoid the loss of.

scale (*n*) system of units of measurement, arranged in order of magnitude; proportion between the real size of an object or area and its representation on paper (e.g. a map or plan).
(*adj*) *large-scale:* extensive, using large resources.

scarcely (see PART II (General) under *Degree*).

scarcely ever (see PART II (General) under *Frequency*).

scatter (*v*) to distribute in various directions.
(*n*) result of scattering.

scope (*n*) range of action, extent.

-scope (see APPENDIX A).

section (*n*) one of a number of separable parts which can be assembled to make a whole; separate group or subdivision within an organization; very thin sample cut from anything and prepared for microscopic examination (also *cross-section:* drawing or photograph of an object which shows the arrangement of its internal parts).

sector (see APPENDIX D under *Shapes, etc*).

seldom (see PART II (General) under *Frequency*).

select (*v*) to choose with care, to choose as most suitable.

self- (see APPENDIX A).

semi- (see APPENDIX A).

sensitive (*adj*) able to react to very slight changes or stimuli.

separate (*v*) to come apart or cause to come apart or be divided.
(*adj*) by itself, not connected or joined with.

sequence (*n*) order in which objects or events follow one another in time and space.

series (*n*) a number of similar or related objects or events occurring or arranged in a sequence (q.v.).

set (*n*) a number of similar or related objects or quantities grouped together for a certain purpose.
(**v*) to put in a certain position; (of instruments, etc.) to adjust or make ready for use; *set up:* to establish, to organize, to equip.

shallow (*adj*) not deep (q.v.).

shape (*n*) form (q.v.), outline.

sharp (*adj*) marked, sudden, intense.

sheet (*n*) thin, flat and extensive area of a substance.

shift (*v*) to move, to change position.
(*n*) a change of position.

shock (*n*) strong, sudden disturbance.

short (*adj*) opposite of long (q.v.).

shorten (*v*) to make short.

show (*v*) to cause to be seen; to prove, to demonstrate.

side (see PART II (General) under *Position*).

sideways (see PART II (General) under *Movement*).

sign (*n*) mark used to represent concepts or mathematical operations (see also APPENDIX C); something which indicates the existence of proof or evidence.

significance (*n*) special meaning or importance in the given conditions.

significant (*adj*) having special meaning or importance in the given conditions.

similar (*adj*) having nearly the same characteristics.

similarly (see PART II (General) under *Others*).

simple (*adj*) opposite of complex (q.v.).

simply (see PART II (General) under *Restriction*).

since (see PART II (General) under *Time* and (Modifying Connectives) under *Result*).

single (*adj*) having or being no more than one.

sink (**v*) to move slowly downwards or below the level of a surface.

-sion (see APPENDIX A).

-sis (see APPENDIX A).

size (*n*) degree of largeness or smallness.

skill (*n*) ability (q.v.).

skin (*n*) (see APPENDIX D under *Parts of the Human Body*).

slanting (*adj*) oblique (q.v.).

slight (*adj*) (see PART II (General) under *Degree*).

slightly (see PART II (General) under *Degree*).

slope (*n*) a surface at an angle between horizontal and vertical.
(*v*) to be at an angle between horizontal and vertical.

slow (*adj*) opposite of fast (q.v.).

smooth (*adj*) opposite of rough (q.v.).

so as to (see PART II (General) under *Others*).

so far (see PART II (General) under *Time*).

soft (*adj*) opposite of hard (q.v.).

soil (*n*) layer of earth in which plants, trees, etc. grow.

solely (see PART II (General) under *Restriction*).

solid (*n & adj*) state of matter which is neither liquid nor gas, and which has a definite shape and dimensions.

soluble (*adj*) able to be dissolved (q.v.).

solution (*n*) act or process of solving (q.v.); complete mixture of molecules of two or more different substances, especially dissolved in a liquid.

solve (*v*) to find the right answer to a problem.

some (see PART II (General) under *Number and Quantity*).

sometimes (see PART II (General) under *Frequency*).

somewhat (see PART II (General) under *Degree*).

as **soon as** (see PART II (General) under *Time*).

sort (*n*) kind (q.v.).
(*v*) to arrange according to certain characteristics, to group (q.v.).

source (*n*) origin (q.v.).

space (*n*) distance, area or volume between bodies; (*outer*) *space:* region beyond the earth's atmosphere.
(*v*) to arrange objects or events with definite intervals of time or space between them.

special (*adj*) not ordinary, for a definite and exclusive purpose.

species (*n*) group of living organisms which have similar characteristics and are able to interbreed.

specific (*adj*) precise, strictly defined and limited, not general.

specimen (*n*) an individual sample (q.v.) usually used for intensive study and analysis.

speed (*n*) rate of movement measured in terms of unit length per unit time.
(*v*) (usually *speed up*) to cause to go faster, to accelerate.

in **spite of** (see PART II (Modifying Connectives) under *Modification*).

split up into (**v*) to divide into two or more smaller parts.

spot (*n*) place (q.v.); area, usually relatively small and roughly circular, of a different colour from the rest.
(*v*) to detect (q.v.) something which is difficult to see.

square (*n & adj*) (see APPENDIX D under *Shapes, etc.*); (math.) the result of multiplying a number by itself, once.
(*v*) to multiply a number by itself, once.

stability (*n*) quality of being stable.

stable (*adj*) fixed, not liable to alter or vary; (chem.) which does not easily break down into its component parts.

stage (*n*) a relatively well-defined section (q.v.) in a process of development, phase (q.v.).

stand (**v*) to resist; to be, or be put in a position, especially vertical; *to stand for:* to represent (q.v.).

standard (*n*) accepted unit or description forming a basis for the measurement of similar things.

standardize (*v*) to apply a common standard to.

start (*v*) to begin.
(*n*) a beginning.

state (*n*) manner in which a thing exists; phase or stage (q.v.); (soc.) the government, in general, an imprecise, misleading and dangerous pseudo-concept.
(*v*) to say or explain in clear and definite terms.

steady (*adj*) regular in movement, speed, etc.; constant, unchanging.

steep (*adj*) sloping at a sharp angle, sudden, marked.

stem from (*v*) to have its origin in.

step (*n*) a stage (q.v.); one of a series of established actions having a common objective.
(*v*) *to step up:* to cause to increase.

still (see PART II (General) under *Time*).

stimulate (*v*) to cause a stimulus to act.

stimulus (*n*) that which causes activity or which produces an answering action in the object on which it acts.

store (*v*) to collect and keep for future use; to accumulate (q.v.).
(*n*) that which is stored.

straight (*adj*) not curved or bent (q.v.).

straighten (*v*) to make straight.

strain (*n*) (eng.) change in size or shape as a result of a load (q.v.) acting on it; (med.) damage caused by over-extending a muscle or the nervous system.

stream (*n*) steady flow.

strength (*n*) capacity to exert or resist force.

stress (*n*) (eng.) force associated with a strain (q.v.) measured by dividing the load (in units of weight) by the area over which it operates, tension; (med.) nervous strain.
(*v*) to indicate as very important, to emphasize.

strong (*adj*) having strength (q.v.); marked, having a high degree of intensity.

structure (*n*) that which is made up of a number of inter-related parts to form a complex whole; special pattern formed by the arrangement of component parts.

sub- (see APPENDIX A).

subject (*n*) that which is spoken about or studied; that on which an experiment is performed.
(*v*) to cause to undergo (q.v.).

subject to (see PART II (Modifying Connectives) under *Condition*).

subjective (*adj*) influenced by factors other than the evidence, e.g. personal feelings.

subsequently (see PART II (Modifying Connectives) under *Order of Events*).

substance (*n*) matter (q.v.) in unspecified form, material (q.v.).

substantially (see PART II (General) under *Degree*).

subtract (*v*) to take away from.

succeed (*v*) to attain the objective aimed at.

success (*n*) result of succeeding.

successful (*adj*) having success; obtaining very good results.

successive (*adj*) coming one after the other.

such as (see PART II (General) under *Others*).

sufficient (*adj*) being or having the number or amount required, enough, adequate.

suggest (*v*) to cause to be considered as a possibility.

suitable (*adj*) having the characteristics required, fit.

sum up (*v*) to give a short account, including only the most important points.

super- (see APPENDIX A).

supply (*n*) amount available.
(*v*) to provide (q.v.).

support (*v*) to resist the weight, pressure, etc. of; to strengthen (conclusions, theories, etc.).
(*n*) that which supports.

supra- (see APPENDIX A).

surely (see PART II (Modifying Connectives) under *Emphasis*).

surface (*n*) extension in two dimensions only; top layer of a liquid or solid.

surplus (*n*) excess (q.v.), especially of goods or money.

syn(m)- (see APPENDIX A).

synthesis (*n*) act or result of combining separate parts or elements to form a complex whole or system.

synthetic (*adj*) belonging to or resulting from a process of synthesis (q.v.); not natural (q.v.).

system (*n*) group of inter-related objects or parts working together to form a whole; set of ideas, theories, measurements, etc. arranged in an orderly way to form a unity.

T

tackle (*v*) to deal with, especially energetically.

take place (**v*) to happen, to occur.

tap (*v*) to exploit (q.v.).

team (*n*) a group (especially of scientists) working on a common problem.

technical (*adj*) belonging to a specific branch of knowledge; belonging to specific mechanical or industrial processes.

technician (*n*) person skilled in operating a machine or carrying out a specific technological process.

technique (*n*) special skill(s) or method(s) used to achieve a particular scientific or industrial purpose.

technology (*n*) branch of knowledge concerned with the application of science to mechanical, agricultural or industrial processes.

temporary (*adj*) intended to last for a short time only.

tend (*v*) to have a tendency (q.v.) towards.

tendency (*n*) state of being liable to develop in a certain manner or direction; trend (q.v.).

tension (*n*) stress (q.v.).

tentative (*adj*) suggested or done as a trial (q.v.) only; provisional.

term (*n*) word or symbol used to indicate a specialized concept; long-term, (opposite = short-term): see under **long;** *in terms of:* as, from the point of view of.
(*v*) to give a name to.

test (*n*) process or examination for the purpose of establishing the characteristics or behaviour of something before it is used freely.
(*v*) to subject to a test or series of tests.

tetra- (see APPENDIX A).

that is (to say) (see PART II (General) under *Others*).

then (see PART II (General) under *Time* and (Modifying Connectives) under *Result* and *Order of Events*).

theoretical (*adj*) having to do with theory or pure science (often opposed to practical (q.v.)).

theory (*n*) a set of inter-related concepts put forward to account for observed facts or phenomena.

there (see PART II (General) under *Position*).

thereafter (see PART II (General) under *Time*).

therefore (see PART II (Modifying Connectives) under *Result*).

therm- (see APPENDIX A).

thick (*adj*) having relatively great width or depth; not flowing easily; dense (q.v.).

thin (*adj*) opposite of **thick.**

through (see PART II (General) under *Movement* and under *Others*).

throughout (see PART II (General) under *Position* and *Time*).

thus (see PART II (Modifying Connectives) under *Result*).

tight (*adj*) fixed, not able to move freely, fitting very closely.

till (see PART II (General) under *Time*).

tilt (*v*) to cause to be at an angle, slanting or sloping.

time (*n*) distance between events, measured in terms of past, present, and future; *times* (see PART II (General) under *Number and Quantity* and *Frequency*); *in time* (see PART II (General) under *Time* and (Modifying Connectives) under *Order of Events*); *at times* (see PART II (General) under *Frequency*); *from time to time* (see PART II (General) under *Frequency*).
(*v*) to measure the time taken by.

tiny (*adj*) very small, minute (q.v.).

-tion (see APPENDIX A).

tongue (*n*) (see APPENDIX D under *Parts of the Human Body*).

tool (*n*) instrument or machine used for a practical purpose.

top (*n*) the highest point or upper surface (see also PART II (General) under *Position*).

total (*n*) amount reached by all the items added together.
(*adj*) of a total; complete.

totally (see PART II (General) under *Degree*).

towards (see PART II (General) under *Position*).

trace (*n*) very small amount of; sign (q.v.); record made (usually on graph paper) by an automatic instrument.
(*v*) to detect (q.v.).

track (*n*) signs (usually continuous) left by an object moving over a surface or through a medium.
(*v*) to keep a continuous record of the position of.

train (*v*) to subject to a process through whch skills are obtained.

training (*n*) process of being trained.

transfer (*v*) to cause to change position from one object to another.
(*n*) result or process of transferring.

transmit (*v*) to send (e.g. messages, genetic characteristics, sound waves, etc.) from one object or place to another.

travel (*v*) to move, to pass from one place to another.

treat (*v*) to apply a given process to.

treatment (*n*) process of treating.

treble (*v*) to multiply by three; to make or be made three times as big, etc.
(*adj*) three times the amount; compound, consisting of three identical items;

trend (*n*) general direction of development, tendency.

tri- (see APPENDIX A).

trial (*n & adj*) test (q.v.).

true (*adj*) in accordance with all the evidence; (eng.) fitting accurately, in line.

try (*v*) to attempt (q.v.).

turn (*v*) to change direction; to revolve or cause to revolve (q.v.); to change from one state, colour, etc. to another (often with 'into'); *turn out:* to be identified eventually as; to be found or discovered. (*n*) act of turning (q.v.); *in turn:* as a result.

twice (see PART II (General) under *Frequency*).

-ty (see APPENDIX A).

type (*n*) kind, class (q.v.).

typical (*adj*) representing a type; characteristic (q.v.).

U **ultimately** (see PART II (Modifying Connectives) under *Order of Events*).

ultra- (see APPENDIX A).

uni- (see APPENDIX A).

under- (see APPENDIX A).

under (see PART II (General) under *Position* and *Number and Quantity*).

undergo (*v) to be subject to the action and effects of.

undoubtedly (see PART II (General) under *Degree*).

uni- (see APPENDIX A).

uniform (*adj*) having the same given characteristics, not varying.

unit (*n*) one of a class or set (q.v.); quantity or amount used as a standard of measurement.

unite (*v*) to bring together to form a whole.

unless (see PART II (Modifying Connectives) under *Condition*).

until (see PART II (General) under *Time*).

upkeep (*n*) maintenance (q.v.).

upon (see PART II (General) under *Position*).

upper (*adj*) being above (of two).

upright (*adj*) vertical (q.v.) (see also PART II (General) under *Position*).

upside down (see PART II (General) under *Position*).

up to (see PART II (General) under *Number and Quantity* and *Time*).

upwards (see PART II (General) under *Movement*).

use (*n*) the purpose of something, the way in which it is of value, the work a thing is able to do.
(*v*) to make a use of, to employ for a purpose; *use up:* to use until nothing remains or until the supply is exhausted.

useful (*adj*) of use, producing good results.

usual (*adj*) which happens on most occasions out of a given number of occasions.

usually (see PART II (General) under *Frequency*).

utilize (*v*) to put into action, to make advantageous use of.

V **value** (*n*) measure of what a thing is worth, or of its usefulness; magnitude, intensity.

valid (*adj*) suggested by evidence, true; that applies equally to, that includes.

validity (*n*) quality of being valid.

variation (*n*) fluctuation, quality of varying (q.v.).

variety (*n*) type, kind; (biol.) subdivision of a species (q.v.); quantity of different types or kinds.

vary (*v*) to become, or cause to become different from the norm (q.v.); to change.

velocity (*n*) speed (q.v.); (phys.) speed in a certain direction.

verify (*v*) to test the truth or accuracy of.

vertical (*adj*) at an angle of 90° to the horizontal (q.v.). (see also PART II (General) under *Position*).

view (*n*) that which is seen; an estimate (q.v.), an opinion; *with a view to:* with the purpose of; *in view of* (see PART II (Modifying Connectives under *Result*).

violet (see APPENDIX D under *Colours*).

visible (*adj*) which can be seen or observed.

volume (*n*) capacity (q.v.); usually measured in cubic units (e.g. c.c., cu. inches, etc.); amount of.

W

want (*n*) a lack, a deficiency (q.v.); a need.
(*v*) to have need for.

waste (*n*) loss caused by bad use or by lack of use.
(*v*) to cause waste.

wave (*n*) a form of transmission of energy in which the particles, molecules or fields vary periodically both in time and space.

weak (*adj*) opposite of **strong** (q.v.).

weakness (*n*) quality of being weak.

weigh (*v*) to measure weight.

weight (*n*) amount of force with which a body is attracted to the centre of the earth, usually measured in terms of kilogrammes or pounds; relative importance or value.

when (see PART II (General) under *Time*).

where (see PART II (General) under *Position*).

whereas (see PART II (Modifying Connectives) under *Modification*).

whether (see PART II (Modifying Connectives) under *Condition*).

while (whilst) (see PART II (General) under *Time* and (Modifying Connectives) under *Modification*).

white (see APPENDIX D under *Colours*).

whole (*n*) all the parts considered together; a thing complete in itself.
(*adj*) complete.

wholly (*adj*) (see PART II (General) under *Degree*).

wide (*adj*) having width (q.v.); having more width than usual or expected; extensive.

width (*n*) measurement from side to side.

within (see PART II (General) under *Time*).

work (*n*) function (q.v.); (phys.) measure of the transfer of energy expressed in calories, ergs, foot-pounds, etc.

work out (*v*) to obtain gradually, over a period; to give a satisfactory result.

wrong (*adj*) opposite of **right** (q.v.).

X

x-rays (*n*) electromagnetic waves with very short wavelengths which can pass through substances through which light is unable to pass.

Y

yellow (see APPENDIX D under *Colours*).

yet (see PART II (General) under *Time* and (Modifying Connectives) under *Modification*).

yield (*n*) amount produced, output (q.v.).
(*v*) to produce.

young (*adj*) in the early stages of growth.
(*n*) living organisms (specially of higher forms of life) which are in the early stages of growth.

Z

zone (*n*) region (q.v.); area distinguished by certain special characteristics.

Part II Structural Words and Phrases

(In the following lists and corresponding illustrative sentences, the main structural words and phrases met with in scientific English have been grouped as far as possible according to their functions. Hence the same English word or phrase may occur in more than one place, and may therefore have different equivalents in the vernacular.)

1 GENERAL STRUCTURAL WORDS AND PHRASES

INDICATING POSITION

above The air *above* industrialized areas tends to be heavily polluted. (See also **over**.)

after An exclamation mark placed *after* a number (e.g. 5!) indicates a number obtained by multiplying together all the integers between it and 1 (i.e. $1 \times 2 \times 3 \times 4 \times 5 = 120$).

against In an internal combustion engine, a piston compresses a gas *against* the top part of the cylinder.
In an experiment, it is always advisable to measure the performance of the subjects *against* the performance of one or more controls.

all over The problem of underdevelopment is one which concerns governments *all over* the world.

among *Among* the many subjects discussed at the international conference, one of the most important concerned the relation of the scientist to society.

around The rings *around* the planet Saturn are apparently made up of ice crystals. (See also under *Number and Quantity*.)

away The nearest star is 10^{14} km *away*. (See also under *Movement*.)

back The engines of many present-day jet aircraft are at the *back* of the plane in order to reduce cabin noise and vibration. (See also under *Movement*.)

before Minus and plus signs are placed *before* the numbers to which they refer. (See also **front**; see also under *Time*.)

below Absolute zero (0°K) is approximately 273° *below* zero. (See also **under** and **beneath**.)

beneath The layer of soil immediately *beneath* (below) the surface or topsoil is called the B horizon.

beside In a motor-car engine, the cylinders are usually arranged in a line, one *beside* the other.

between A geological period of warm climate occurring *between* two glacial periods is called an interglacial.

beyond Radio telescopes have been able to probe space *beyond* the range of ordinary optical telescopes. (See also under *Movement*.)

bottom Currents of cold water tend to flow near the *bottom* of lakes and oceans.

front Shock waves are produced when a body moving at hypersonic speeds compresses the air in *front* of it.

here A famous philosopher used to say: 'As I am *here*, in this place, I cannot be *there* in that other place, since a body cannot occupy two different places at the same time.'

horizontal Sedimentary rocks are generally laid down (deposited) in *horizontal* layers, which may subsequently be tilted by pressures from any direction.

in Zoological specimens are usually preserved *in* bottles containing alcohol or formalin.

in front of (See *front*.)

inner The trunk of a tree is formed by an *outer* layer of bark and an *inner* structure of wood.

inside The pressures and tensions *inside* the earth give rise to such surface phenomena as vulcanism (volcanism, U.S.) and earthquakes.

inside out The concept of turning 3-dimensional bodies *inside out* (i.e. with the original inner surface on the outside) is one of the interesting problems of topology.

near The zones having the heaviest precipitation (rainfall) in the world are on or *near* the equator.

oblique Straight lines or surfaces which are neither horizontal nor vertical are called *oblique* or *slanting*.

on Much inefficiency in farming (agriculture) is due to growing crops *on* the wrong sort of soil.

on top of When examining specimens with the biological microscope, a coverslip is usually placed *on top of* the slide.

opposite In a motor-car engine, the cylinders are usually arranged in two rows *opposite* (facing) one another.

out If the front and back wheels of a vehicle are not directly behind each other, they are said to be *out* of alignment (out of line).

outer (See **inner**.)

outside Artificial satellites circle the earth *outside* its atmosphere.

over The air *over* large bodies of ice tends to be cooled to the point where almost continuous precipitation occurs. (See also **above**; see also under *Number and Quantity*.)

overhead The zenith is the point in the heavens directly *overhead* of the observer.

side An engineering diagram is usually drawn from two points of view, i.e. as seen from above (the plan) and as a *side* view (elevation or profile).
(For *at the side of*, see **beside**.)

slanting (See **oblique**.)

there (See **here**.)

throughout (See **all over**; see also under *Time*.)

top In geology, the *top* layer (stratum) of rock is not always the most recent, as might be expected, owing to the fact that erosion may have uncovered rocks deposited in earlier times.

under (See **below** and **beneath**; see also under *Number and Quantity*.) Also: in the production of rice, the fields are flooded so that the seed grows *under* the water.

upright When reading a thermometer, it is essential to hold the instrument in an *upright* (*vertical*) position, otherwise the reading may be wrong.

vertical (See **upright**.)

upside down Images seen through the lens of various optical instruments, such as a camera, appear *upside down*, i.e. the tops of the objects appear at the bottom of the image, and *vice versa* (the reverse).

where The place *where* experiments are generally performed is the laboratory.

INDICATING MOVEMENT

across Television (TV) pictures can be transmitted *across* the Atlantic by means of artificial satellites.

along Messages from the brain are sent to the extremities (ends) of the body *along* the nerves in a fraction of a second.

as far as According to reliable sources the sound of the volcanic explosion of Krakatao, in Indonesia, was heard in places *as far as* 300 miles away.

back The flying animals called bats avoid obstacles by emitting sound signals which then bounce *back* from the obstacles and are perceived by the bat.

backward(s) When a radioactive nucleus emits a particle, the reaction causes the nucleus to move *backwards* in a direction opposite to that of the particle.

beyond No space probe has yet reached *beyond* the planets Mars and Venus.

down Loose soil tends to move *down* slopes under the influence of gravity, thus causing denudation of the upper slopes.

downward(s) Heavy rain falling on unprotected soil carries the minerals *downwards* beyond the reach of the plant roots; this process is known as 'leaching'.

forward(s) When a horizontal bar AB is struck with a hammer at the end 'A', the resulting shock waves move *forwards* along the bar until they reach the other end 'B'.

into Alcohol absorbed *into* the bloodstream tends to slow down the speed of nervous reactions.

inward(s) When a volcano erupts and large volumes of lava are ejected, this may cause a large cavity to be formed immediately below it. The cone of the volcano may then collapse *inwards*, forming what is known as a 'caldera'.

outward(s) In large bodies of ice, such as the Antarctic or Greenland ice-caps, the ice moves *outwards*, from the centre to the margins, under the influence of pressure ('plastic flow').

sideways The needle of a magnetic compass is deflected *sideways* if a piece of iron is placed on one side of it.

through When a strong solution is separated from a weak solution by a semi-permeable wall (e.g. a cell-wall), the former passes *through* the wall, from one side to the other, until equilibrium is reached ('osmosis').

toward(s) Convection currents cause polar waters to move *towards* the equator.

upwards The layers of a liquid or gas in contact with a source of heat expand and become less dense. They rise *upwards* and their place is taken by colder and denser layers moving downwards ('convection').

INDICATING NUMBER AND QUANTITY

about Radio waves are reflected by an upper layer (the 'Heaviside layer') of the earth's atmosphere, which is at an altitude of *about* (approximately, around) 60–80 miles.

any In an equilateral triangle, *any* side and *any* angle is equal to either of the other two.
(For *not any*, see **none**.)

approximately (See **about**.)

around (See **about**.)

a few The female salmon produces about 5000 eggs; of this number only *a few*—approximately 1 out of 10—become adult fish (*a few* = some, although not very many, i.e. a positive idea. cf. *few*).

few *Few* countries in the world at the moment are able to support the heavy costs of space research programmes (few = a very limited number, i.e. a negative idea. cf. *a few*).

less than In tropical areas, deserts exist where the net annual precipitation is *less than* (under) 5″ (i.e. 125 mm.)

a lot of (See **many** and **much**.)

many Although *many* (a lot of) countries have started a programme of industrialization, only a few are able to provide a full range of manufactured goods for their people.

more than Desert zones may occur where the net annual precipitation is *more than* (over) 5″ (up to a maximum of 20″), if the precipitation is concentrated on a single or only a few occasions during the year.

much One of the problems of internal combustion engineering is that *much* (a lot) of the energy generated cannot be converted into useful work, and is therefore wasted.

none There are several species of land mammals in the north Polar regions but *none* (not any) in the south Polar regions.

only Some elementary particles—e.g. various types of mesons—have lifetimes of *only* 10^{-8} to 10^{-6} seconds.

over (See **more than.**)

some *Some*—but not many—species of fish generate a weak electric field in order to detect objects in the water.

times The most famous physical equation of modern *times*—$E = mc^2$—means that energy equals mass *times* (i.e. multiplied by) the square of the speed of light.

under (See **less than.**)

up to Pressures *up to* 130,000 atmospheres have been produced under laboratory conditions.

INDICATING TIME

in **advance** Under modern conditions it is often difficult to make an accurate estimate *in advance* (beforehand) of the cost of a complex production programme.

after The efficiency of a new fertilizer in agriculture can only be assessed *after* a large number of experiments under different conditions have been made.

afterwards In an electrical storm, the discharge (i.e. the lightning) is seen first; the noise of the discharge (i.e. the thunder) is heard some time *afterwards*.

already Although the laws relating to air pressure were not formulated until the seventeenth century, air pumps had *already* been used for many centuries previously.

as soon as The new electronic computer was put into production *as soon as* (when) it had completed its testing programme.

at Very few scientists were able to see the practical implications of the Einstein relation ($E = mc^2$) *at* the time it was first formulated.

at present (See under **present.**)

before It was difficult to investigate the behaviour of elementary particles *before* high-energy particle accelerators were developed.

beforehand (See **in advance.**)

by A manned space-craft is expected to land on the moon *by* 1970.

during The work of molecular biologists *during* (over) the last 10 years has thrown light on the complex mechanisms of heredity.

eventually In 1798 Malthus predicted that, at the current rates of growth, food supplies would *eventually* (in time) be insufficient for the population.

hitherto The traditional methods of shaping metals and alloys have *hitherto* (up to the present) been by means of expensive and short-lasting cutting tools. In the future, however, laser beams or electrical discharges will probably be widely used instead.

in (See **more than.**)

just The first trial run of the new machine lasted from 10.00 hours to 11.00 hours. It is now 11.01, so the test has *just* finished.

in the **meantime** Modern engineering requires new materials with special qualities which are not always available at present, although future research may develop them; *in the meantime*, improvements in traditional materials are being tried.

more than Although problems of abnormal psychology have been studied for *more than* (over) 300 years, it is only *within* (in) the last 50 years that marked advances in treatment have been made.

now Many hitherto non-industrialized countries are *now* stepping up their efforts to increase their production of manufactured goods.

nowadays In the past, natural resources were thought to be unlimited; *nowadays*, it is realized that many items are likely to be used up in the comparatively near future.

once The precise dating of geological and anthropological specimens was possible for the first time, *once* the rate of decay of radio-active elements had been accurately determined. (See also **as soon as.**)

over (See **more than**; see also **during**.)

at **present** There are no known cures for degenerative diseases (i.e. diseases of old age) *at present*, but many approaches to the problem are being investigated.

recently At the end of 1957 it could be said that man had only *recently* succeeded in sending a vehicle into space.

since In its beginning stages, psychology was considered to be mainly an observational science; *since* the 1920s, however, considerable emphasis has been placed upon carefully-controlled experimentation.

so far (See **yet.**)

still Urbanization (the process of living in cities) is so recent in human society that even the most urbanized countries *still* show the rural origins of their institutions.

then In scientific research, ideas are usually first developed in the laboratory, and are *then* applied by technologists for practical use. (See also under 'Modifying Connectives', *Order of Events*).

in **time** (See **eventually**.) Also: Many people believe that in view of the present population explosion and shortage of natural resources, birth control methods will not be employed *in time* to prevent a major world disaster.

thereafter Economic growth in developing countries is usually slow and difficult until a certain point (the so-called 'take-off' stage) is reached; *thereafter* (from this point onwards). it tends to become relatively rapid and easy.

throughout The possibility of variations in external factors must constantly be kept in mind *throughout* an experiment or series of experiments, from beginning to end.

till (See **until.**)

until Only three types of elementary particle were thought to constitute the 'building blocks, of matter *until* (till, up to) 1947. Research since this date has revealed the existence of many more.

up to (See **until.**)

when (See **as soon as.**)

while The conditions which cause a given phenomenon may change *while* the phenomenon is being subjected to observation or experiment. Control groups are therefore necessary.

within (See **in.**)

yet In spite of the enormous amount of damage caused annually by earthquakes, no system of prediction or early warning has *yet* been developed.

INDICATING FREQUENCY

always A magnetic needle *always* points to the magnetic north unless disturbed by external factors.

frequently Owing to the difficulty at present of predicting the weather conditions over long periods of time, estimates of agricultural production are *frequently* (often) highly inaccurate.

generally Black-and-white photographs are *generally* (usually) used in air-mapping, although colour photographs are also used for certain special purposes, e.g. studies of vegetation.

hardly ever In spite of the widespread occurrence of volcanoes, volcanic energy is *hardly ever* (scarcely ever) used as a source of power.

never Up to now, atomic energy has *never* been used for aircraft propulsion.

occasionally Radioisotopes are generally used nowadays for accurate absolute dating of geological and archaeological specimens, though various other methods are *occasionally* used (used from time to time) for this purpose.

often (See **frequently**.)

once The results of the investigation were not conclusive since the crucial experiment had been performed only *once* and had not been repeated. (See also under *Time*.)

rarely The wavelengths of light are *rarely* (seldom) given in centimetres or millimetres, as radio waves are, but are almost always expressed in Angstrom units ($Å = 10^{-10}$ m.).

scarcely ever (See **hardly ever**.)

seldom (See **rarely**.)

sometimes Progress in a given line of research *sometimes* (at times), but not always, depends on the development of new types of instruments.

from **time** to **time** (See **occasionally**.)

times (See **twice** and also under *Number and Quantity*.)

at **times** (See **sometimes**.)

twice Crucial experiments must be performed not merely once, or even *twice* (two times), but as many *times* as is considered necessary to convince the scientific community of the validity of the results.

usually (See **generally**.)

A SCALE OF FREQUENCY

Students may find the following rough scale of frequency useful as a quick aid:

%	
100	always
90	almost always
80	generally, usually
70 60	often, frequently
50	as often as not
40	
30	
20	sometimes, at times
10	occasionally, from time to time, not often
5	rarely, seldom, hardly ever
0	never

INDICATING DEGREE

almost The shape of the earth is *almost* (nearly) spherical, being very slightly flattened at the poles.

comparatively Unlike chemistry, which has been studied systematically for at least 5 centuries, the study of genetics dates only from the beginning of the present century, and is therefore a *comparatively* (relatively) recent branch of science.

completely The implications of the latest anthropological discoveries (as at 1965) have been so far-reaching that previous views about the evolution of man have had to be *completely* (totally, entirely, wholly) revised.

easily It is very difficult to make direct visual observations of the inner organs of the living human body, but they can *easily* (readily) be seen by indirect methods, e.g. X-rays.

enough Although the density of the atmosphere is very low at altitudes above (exceeding) 50 miles, it is still dense *enough* to cause small meteorites to burn up by friction against the air molecules.

entirely (See **completely**.)

especially Although an advanced knowledge of mathematics is required by all types of scientists, it is *especially* (particularly) necessary for engineers and physicists.

exactly One of the initial difficulties in research is to determine *exactly* (precisely) and without ambiguity the nature and limits of the problem to be investigated.

exceedingly Afforestation (the planting of trees) is an *exceedingly* (extremely, highly) effective method of preventing erosion on unproductive hillsides.

extensively (See **largely**.)

to a large **extent** (See **largely**.)

extremely (See **exceedingly**.)

fairly Many of the conditions obtaining in space can be reproduced in the laboratory *fairly* well; there are, however, some factors which are impossible to duplicate on earth.

fully Chemical anaesthetics may leave a subject confused or semi-conscious for an hour or more before and after an operation; a new type of electrical anaesthesia is being developed which can cause almost immediate unconsciousness and, after the operation, a rapid return to a *fully* (completely) conscious state.

hardly At present, the known supplies of certain minerals are *hardly* (scarcely) sufficient for current needs, let alone (i.e. without considering) future requirements.

highly (See **exceedingly**.)

largely Contemporary technology is based *largely* (mainly, primarily, extensively, to a large or considerable extent or degree) on the use of fuels of organic origin.

at (the) **least** Although nothing is known directly about the temperature of the centre of the earth, geophysicists estimate that it must be *at least* 2000–3000°C, probably higher.

be **likely** Chemical methods of controlling pests which cause damage to crops often have undesirable side-effects; for this reason it is *likely* (probable) that biological methods will be applied increasingly in the future.

little In spite of the ever-increasing exploitation of natural resources, which has now reached dangerous proportions, *little* has been done on a world-wide scale to slow down or stop this process. (little = not much and therefore has a negative meaning. cf. *a little*.)

a little In spite of the very great difficulties involved in the study of the mechanisms of heredity, *a little* progress is now being made, and a great deal more can be expected in the near future. (a little = some and therefore has a positive meaning. cf. *little*.)

mainly (See **largely**.)

maybe (See **perhaps**.)

moderately In spite of its comparatively high latitude, the climate of the south of England is only *moderately* cold, due to the influence of warm ocean currents.

at (the) **most** The food and textile industries, although highly mechanized, still rely on the human senses for certain essential operations, such as the tasting of samples and the grading of fibres. Mechanical processes intended to replace these can only achieve equivalent results, *at (the) most*, whilst having the disadvantages of being more expensive and less flexible in use.

nearly (See **almost**.)

partially Agricultural production can only *partially* (partly) be increased by means of improvements in farming techniques; many other factors, such as marketing methods, systems of land ownership, etc., also have a large effect on efficiency.

partly (See **partially**.)

particularly (See **especially**.)

perfectly In some cases it is *perfectly* (completely) possible to decide whether a theory is valid or not by performing only one experiment, the so-called 'crucial' experiment.

perhaps *Perhaps* (possibly, maybe) the easiest short-term solution to the problem of undernourishment will be the synthesis of proteins.

possibly (See **perhaps**.)

precisely (See **exactly**.)

primarily (See **largely**.)

probably (See **likely**.)

profoundly A substantial body of evidence supports the common observation that the experiences of very early life *profoundly* (strongly) influence the structure of the adult personality.

quite It is *quite* (completely) impossible for a body to attain speeds faster than that of light.

The number of elements which make up organic compounds is *quite* (moderately, fairly) restricted, although the number of combinations into which they can enter is very large.

rather (See **fairly**; see also **somewhat**.)

readily (See **easily**.)

relatively (See **comparatively**.)

roughly (See **approximately**.)

scarcely (See **hardly**.)

slightly Isotopes are forms of an element which differ only *slightly* in mass from the element itself, e.g. natural uranium (U) has a mass of 238, while the most familiar isotope has a mass of 235.

somewhat Copper has a *somewhat* (rather) higher specific gravity than tin—8·95 compared with 7·31.

specially (See **especially**.)

substantially (See **largely**.)

totally (See **completely**.)

undoubtedly Computers are *undoubtedly* replacing human beings in the taking of decisions in large areas of business, war, and politics.

wholly (See **completely**.)

SCALES OF DEGREE

The student may find the following arrangement of 'Degree' words into very rough 'scales' helpful as a quick aid:

(1) Degrees of certainty or doubt: (i.e. in a relative sense)

%	
100	undoubtedly, perfectly, absolutely, clearly, obviously
90	
80	approximately, about, roughly
70	probably, be likely
60	
50	possibly, perhaps, maybe
40	
30	
20	unlikely, improbable
10	
5	be very (highly, most, extremely) improbable, etc.
0	

(2) Degree in an absolute sense:

%	
100	completely, fully, entirely, exactly, precisely, quite[1]
90	extremely, exceedingly, profoundly, highly
80	nearly, almost
70	very, strongly, largely, to a large extent, etc., mainly
60	
50	fairly, comparatively, relatively, moderately, quite[1];

[1] *Quite* is an ambiguous word—see examples.

40
30 partially, partly
20
10 somewhat, rather, a little, a bit, not very
5 slightly, weakly
0

NOTE: The words in the above scales have been given in their adverbial form; the corresponding adjectival form is, however, also frequently used.

INDICATING A RESTRICTION

exclusively Published papers need not always present the results of original research; they may sometimes consist *exclusively* (only, solely, merely, simply) of a synthesis of previous work in the field. These may be very useful, however, in cases where the original material is widely scattered and difficult to obtain.

merely (See **exclusively**.)

only (See **exclusively**.)

simply (See **exclusively**.)

solely (See **exclusively**.)

OTHERS

in **accordance** with All pieces of apparatus should be set up and used strictly *in accordance with* the makers' instructions.

according to *According to* (on the authority of) the special theory of relativity, the velocity of light is the same for all observers, irrespective of their own velocities.

In the periodic table, the chemical elements are arranged *according to* (i.e. in a way that is related with) their increasing atomic weight.

and so on 2, 4, 6, 8 *and so on* (etcetera) are even numbers; 1, 3, 5, 7, *and so on* are odd numbers.

as far as . . . is concerned There are currently two main theories about the origin of the universe: the continuous creation or 'steady-state' theory and the 'Big Bang' (explosion) theory. *As far as* the former *is concerned* (As regards, with regard to the former), the difficulties its acceptance implies seems to outweigh its possible advantages, whereas evidence for the latter theory is in fact increasing.

apparently The results of a series of experiments showed an *apparently* significant periodic fluctuation, but further investigation revealed that this was due to a faulty piece of apparatus.

as follows The so-called even numbers are *as follows*: 2, 4, 6, 8, and so on.

as regards (See **as far as . . . is concerned**.)

consistent with For many centuries it was believed that the sun circled the earth. Although this theory was *consistent with* the facts as then observed, it was, of course, erroneous.

on the **contrary** It is often stated or implied that the scientist is not fitted to play a role in political affairs. It seems to us, however, that *on the contrary*, his special qualifications and responsibilities require his active participation in the process of government.

conversely In a steam-engine, energy in the form of heat is converted into motion; *conversely*, when a meteorite enters the earth's atmosphere, motion is converted into heat.

either . . . or From two of the illustrative sentences given in this section, it can easily be deduced that numbers are *either* odd *or* even.

etcetera (etc.) (See **and so on**.)

for example (e.g.) Several factors, *e.g.* lack of capital, overpopulation, underproduction, etc. may hinder a country from attaining a high standard of living.

on the **grounds** (of) (that) In the past it frequently happened that a valid scientific theory was rejected *on the grounds* that it conflicted with current religious or other dogmatic beliefs.

how Any new item of laboratory equipment is usually accompanied by a technical handbook explaining *how* it has to be installed and operated.

by **means** of Torricelli tested his hypothesis about air pressure *by means of* (through) a simple experiment.

in **order** to Operations (or operational) research—O.R.—was developed during World War II *in order to* (so as to) increase the efficiency of military operations.

with **regard** to (See **as far as . . . is concerned.**)

as **regards** (See **as far as . . . is concerned.**)

in the **same way** A bat locates obstacles in its path by emitting sound waves which the obstacles reflect back to it; *in the same way* (similarly) an aircraft locates obstacles by emitting radio waves which are then reflected back to it.

similarly (See **in the same way.**)

so as to (See **in order to.**)

that is (to say) (i.e.) The rotation of crops, *i.e.* (in other words) the practice of changing the type of crops grown on the same area of soil, is one of the most efficient and cheapest ways of maintaining soil fertility.

through (See **by means of.**)

2. MODIFYING CONNECTIVES[1]

(In the following sections, the two statements given will be linked together by means of the appropriate modifying connective in each case. The degree of emphasis implied is also indicated.)

INTRODUCING ADDITIONAL STATEMENTS

Statement I: The decimal scale is used in science.
Statement II: The binary scale is used in science.

and The decimal scale *and* binary scale are used in science.
(simple added statement—no emphasis)

as well as The binary scale, *as well as* the decimal scale, is used in science.
(emphasizing the binary scale somewhat)

also The decimal scale, and *also* the binary scale, are used in science.
(similar to *as well as*)

apart from *Apart from* (in addition to) the decimal scale, the binary scale is also used in science.
(slightly more emphatic than *as well as*)

besides *Besides* the decimal scale, the binary scale is also used in science.
(similar to *apart from*)

furthermore Besides the decimal scale, the binary scale is also used in science; *furthermore* (moreover), a number of other scales can be used as well, although these are, in practice, used for special purposes (e.g. codes and ciphers).
(usually used when a series of additional statements are made)

in addition (See **apart from.**)

moreover (See **furthermore.**)

INTRODUCING A RESULT

Statement I: The measuring instruments were defective.
Statement II: The experiment was a failure.

accordingly The measuring instruments were defective; *accordingly* (hence, consequently) the experiment was a failure.
(fairly emphatic)

[1] i.e. words and phrases which connect together the various parts of a statement, and at the same time modify it in some way. In this section, as in the previous one (General), they have been grouped approximately according to their modifying function.

as *As* (since, because) the measuring instruments were defective, the experiment was a failure.
(not so emphatic as *accordingly*)

because (See **as.**)

consequently (See **accordingly.**)

due to *Due to* (owing to, in view of, on account of) the fact that the measuring instruments were defective, the experiment was a failure.
(similar to *accordingly*)

hence (See **accordingly**.)

given *Given* the fact that the measuring instruments were defective, the experiment was bound to be a failure.
(inevitable result of the conditions obtaining)

inasmuch as *Inasmuch as* (in so far as, i.e. to the degree that) the measuring instruments were defective, the experiment was a failure; however, some positive results were in fact obtained.
(indicates that the result itself was modified in some way)

in so far as (See **inasmuch as.)**

it follows that The measuring instruments being defective, *it follows that* the experiment was a failure.
(similar to *accordingly*)

on account of (See **due to.**)

owing to (See **due to**.)

since (See **as.**)

then The measuring instruments were defective; the experiment, *then*, was a failure.
(similar to *as*)

therefore The measuring instruments were defective; the experiment was *therefore* (thus) a failure.
(similar to *accordingly*)

INTRODUCING A MODIFICATION

Statement I: One of the methods used for the dating of geological strata is the correlation of fossils.
Statement II: Another method is by measuring the rate of decay of radioactive elements.

alternatively One old-established method for the dating of geological strata is that of the correlation of fossils; *alternatively* (otherwise, on the other hand) the relatively new method of measuring the rate of decay of radioactive elements can be employed in many cases.
(fairly emphatic distinction)

although *Although* (whilst, whereas) the old-established method of dating geological strata by the correlation of fossils is still used extensively, the new method of obtaining accurate absolute dating by measuring the rate of decay of radioactive elements is tending to replace it.
(somewhat more emphatic than *alternatively*)

but The dating of geological strata by means of the correlation of fossils is still used extensively, *but* the new method of obtaining accurate absolute dating by measuring the rate of decay of radioactive elements is tending to replace it.
(less emphatic than *alternatively*)

except *Except* when the conditions are suitable for using the new method of dating geological strata by measuring the rate of decay of radioactive elements, the main method used is that of the correlation of fossils.
(clearly defined limitation or restriction)

however The new method of dating geological strata by measuring the rate of decay of radioactive elements is tending to be used to an increasing extent; *however* (nevertheless), it should be remembered that the old-established method of correlating fossils is also satisfactory for many purposes.
(similar to *although*)

irrespective of *Irrespective of* (notwithstanding, regardless of, in spite of) the fact that the new method of dating geological strata by measuring the rate of decay of radioactive elements is being used to an increasing extent, it should be remembered that the old-established method of correlating fossils is also satisfactory for many purposes.
(slightly more emphatic than *however*)

in spite of (See **irrespective of.**)

nevertheless (See **however.**)

notwithstanding (See **irrespective of.**)

on the other hand (See **alternatively.**)

otherwise (See **alternatively.**)

regardless of (See **irrespective of.**)

whereas (See **although.**)

while (whilst) (See **although.**)

INTRODUCING A CONDITION

Statement I: Present known supplies of organic fuel are likely to be exhausted in the foreseeable future.
Statement II: New sources of power must be exploited on a wide scale.

if Present known supplies of organic fuel are likely to be exhausted in the near future *if* new sources of power are not exploited on a wide scale.
(no particular emphasis)

provided (providing) that *Provided that* (subject to the proviso that) new sources of power are exploited on a wide scale, present supplies of organic fuel are not likely to be exhausted in the foreseeable future.
(more emphatic than *if*)

subject to (See **provided that.**)

unless Present known supplies of organic fuel are likely to be exhausted in the foreseeable future *unless* new sources of power are exploited on a large scale.
(same as *if . . . not*)

whether The possibility that present supplies of organic fuels may be exhausted in the foreseeable future depends on *whether* new sources of power can be exploited on a large scale.
(similar to *provided that*)

INTRODUCING DOUBT OR HYPOTHESIS

Statement I: Under modern conditions, the field of action of the politician seems to be narrowing.
Statement II: In the future, the scientist may take over many of the functions of the politician.

maybe
possibly
perhaps Under modern conditions, the field of action of the politician seems to be narrowing; *maybe* (perhaps, possibly) the scientist will take over many of his functions in the future.

INTRODUCING EMPHASIS

Statement: Some of the factors in successful economic development are non-economic in character: one of these is incentive, or the 'will to develop'.

above all Some of the factors in successful economic development are, *above all* (i.e. more than anything else), non-economic in character . . . etc.
(very strong emphasis on the *non*-economic nature of the factors concerned)

actually Some of the factors in successful economic development are *actually* (in fact, indeed, really) non-economic in character . . . etc.
(emphasizes non-economic nature, and may also imply that the statement is contrary to what is usually believed)

clearly Some of the factors in successful economic development are *clearly* (obviously, certainly, naturally, of course) non-economic in character . . . etc.
('unloaded' emphasis)

certainly (See **clearly**.)

in fact (See **actually.**)

indeed (See **actually.**)

obviously (See **clearly**.)

really (See **actually**.)

surely Some of the factors in successful economic development are, *surely*, non-economic in character . . . etc.
(implies that the writer is trying to persuade the reader of the correctness of the statement—he is asking for the reader's agreement. (NOTE: 'hard' evidence may often be lacking in such cases)

INTRODUCING ORDER OF EVENTS

in the **beginning**
eventually
finally
at first
firstly
initially
lastly
later on
subsequently
next
then
in **time**
ultimately

Theories regarding the shape of the earth have changed throughout the ages. *In the beginning* (at first, at the outset, initially) it was believed that it was flat, although ideas about the exact shape—i.e. whether it was circular, oval, square, rectangular, etc.—varied. *Later on* (subsequently) this concept failed to satisfy some observers and *eventually* (in time) the evidence tending to disprove this idea grew so large that it had to be totally abandoned. An alternative theory was *then* (next) put forward, resting on a number of observed facts, e.g. that the parts of a ship moving away from the observer were seen to disappear below the horizon in sequence—*first*(*ly*) the lower parts, *then* the upper parts and *finally* (lastly) the extreme tops of the masts, until *at last* (ultimately) the ship disappeared from view entirely.

Index of Structures

*Incorporated in the Reading Passage and pointed out in Teacher's Notes—no exercises given.
TN. Reference to page in Teacher's Notes. Presented there as extra material with exercises.

*Incorporated in the Reading Passage and pointed out in Teacher's Notes—no exercises given.
TN. Reference to page in Teacher's Notes. Presented there as extra material with exercises.

*Incorporated in the Reading Passage and pointed out in Teacher's Notes—no exercises given.
TN. Reference to page in Teacher's Notes. Presented there as extra material with exercises.